# CAMBRIDGE LIBRARY COLLECTION

*Books of enduring scholarly value*

## Earth Sciences

In the nineteenth century, geology emerged as a distinct academic discipline. It pointed the way towards the theory of evolution, as scientists including Gideon Mantell, Adam Sedgwick, Charles Lyell and Roderick Murchison began to use the evidence of minerals, rock formations and fossils to demonstrate that the earth was older by millions of years than the conventional, Bible-based wisdom had supposed. They argued convincingly that the climate, flora and fauna of the distant past could be deduced from geological evidence. Volcanic activity, the formation of mountains, and the action of glaciers and rivers, tides and ocean currents also became better understood. This series includes landmark publications by pioneers of the modern earth sciences, who advanced the scientific understanding of our planet and the processes by which it is constantly re-shaped.

## Essay Towards a First Approximation to a Map of Cotidal Lines

Cotidal lines are lines on a map which connect points at which the same tidal level occurs simultaneously. Isaac Newton had explained the movement of the tides by the action of the moon and sun, and Daniel Bernoulli had used Newton's findings to create tide tables for specific locations, but William Whewell wanted to take research further by gathering and analysing information which would link cotidal points or lines across the world. Fellow and eventually Master of Trinity College, Cambridge, Whewell (1794–1866) published this work in 1833. In it he proposes various observations that would need to be undertaken to produce a cotidal map, with detailed descriptions of the factors to be taken into account in computing the results. In 1837, Whewell, several of whose other works are also reissued in this series, was awarded a royal prize medal by the Royal Society for his work on 'tidology'.

# Essay Towards a
# First Approximation to
# a Map of Cotidal Lines

W<small>ILLIAM</small> W<small>HEWELL</small>

CAMBRIDGE
UNIVERSITY PRESS

# CAMBRIDGE
## UNIVERSITY PRESS

University Printing House, Cambridge, CB2 8BS, United Kingdom

Cambridge University Press is part of the University of Cambridge.

It furthers the University's mission by disseminating knowledge in the pursuit of
education, learning and research at the highest international levels of excellence.

www.cambridge.org
Information on this title: www.cambridge.org/9781108073905

© in this compilation Cambridge University Press 2014

This edition first published 1833
This digitally printed version 2014

ISBN 978-1-108-07390-5 Paperback

# ESSAY

## TOWARDS

## A FIRST APPROXIMATION

## TO

# A MAP OF COTIDAL LINES.

———————

BY

THE REV. W. WHEWELL, M.A. F.R.S.

Fellow of Trinity College, Cambridge.

———————

*From the* PHILOSOPHICAL TRANSACTIONS.

———————

## LONDON:

PRINTED BY RICHARD TAYLOR, RED LION COURT, FLEET STREET.

1833.

XI. *Essay towards a First Approximation to a Map of Cotidal Lines. By the Rev.* W. WHEWELL, *M.A. F.R.S. Fellow of Trinity College, Cambridge.*

Read May 2, 1833.

## *Introduction.*

EVER since the time of NEWTON, his explanation of the general phenomena of the tides by means of the action of the moon and the sun has been assented to by all philosophers who have given their attention to the subject. But even up to the present day this general explanation has not been pursued into its results in detail, so as to show its bearing on the special phenomena of particular places,—to connect the actual tides of all the different parts of the world,—and to account for their varieties and seeming anomalies. With regard to this alone, of all the consequences of the law of universal gravitation, the task of bringing the developed theory into comparison with multiplied and extensive observations is still incomplete; we might almost say, is still to be begun.

DANIEL BERNOULLI, in his Prize Dissertation of 1740, deduced from the Newtonian theory certain methods for the construction of tide tables, which agree with the methods still commonly used. More recently LAPLACE turned his attention to this subject; and by treating the tides as a problem of the oscillations rather than of the equilibrium of fluids, undoubtedly introduced the correct view of the real operation of the forces; but it does not appear that in this way he has obtained any consequences to which NEWTON's mode of considering the subject did not lead with equal certainty and greater simplicity; moreover by confounding, in the course of his calculations, the quantities which he designates by $\lambda$ and $\lambda'$, the epochs of the solar and lunar tide (Méc. Cél. vol. ii. p. 232. 291.), he has thrown an obscurity on the most important differences of the tides of different places, as Mr. LUBBOCK has pointed out.

LAPLACE also compared with the theory observations made at Brest from the year 1711 to 1715; and showed that the laws which, according to the theory, ought to regulate the times and heights of the tides, may, in reality, be traced in the averages of this series of observations.   In pursuance of his advice also, a new series of observations was undertaken at the same port, with the intention that it should be continued, at least, during one period of the motion of the nodes of the moon's orbit.   The new observations were begun in 1806, and have since been carried on without interruption.   Of the observations thus made, LAPLACE subjected to a mathematical discussion those for sixteen years, beginning with 1807; and M. BOUVARD, who performed the requisite calculations, employed nearly 6000 tide-observations.   In our own country also, Mr. LUBBOCK has given the results of the examination of about 13,000 tide-observations, made at the London Docks, from 1808 to 1826, in a Memoir recently published in these Transactions.   These results are very important, in consequence of their consistency with theory and with each other; the calculations by which they were obtained were performed by Mr. DESSIOU; and the task which he has thus executed, is, perhaps, in the amount of labour, and in the judicious and systematic mode of its application, not inferior to any of the most remarkable discussions of large masses of astronomical or meteorological observations by other modern calculators.

But in the meantime no one appears to have attempted to trace the nature of the connexion among the tides of different parts of the world.   We are, perhaps, not even yet able to answer decisively the inquiry which BACON suggests to the philosophers of his time, whether the high water extends across the Atlantic so as to affect contemporaneously the shores of America and Africa, or whether it is high on one side of this ocean, when it is low on the other: at any rate such observations have not been extended and generalized.

It will easily be understood that we may draw a line through all the adjacent parts of the ocean which have high water at the same time; for instance, at 1 o'clock on a given day.   We might draw another line through all the places which have high water at 2 o'clock on the same day, and so on. Such lines may be called *cotidal* lines; and they will be the principal subject of the present essay.

It might perhaps be supposed at first that we have now considerable materials for drawing such cotidal lines upon our maps.   The time of the tide has been observed and recorded over a large portion of the earth's surface, by residents or by voyagers, during the last two centuries; and we have in many works tables of the *establishment* of a long list of places. There are, however, in these statements, certain errors and imperfections, which prevent our being able as yet to determine the course of the cotidal lines with accuracy, or even to obtain with certainty a first approximation to these lines. But before we explain the defects of our observations, it will be proper to say a few words on the general properties of the cotidal lines.

The cotidal line for any hour may be considered as representing the summit or ridge of the *tide-wave* at that time; in which expression we mean, by the tide-wave, that protuberance of water upon the surface of the ocean which moves along the seas, and by its motion brings high-water and low-water to any place, at the time when the elevated and the depressed parts of the watery surface reach that place. The cotidal lines for successive hours represent the successive positions of the summit of this wave; and if we suppose a spectator, detached from the earth, to perceive the summit of the wave, he will see it travelling round the earth in the open ocean once in twenty-four hours, accompanied by another at twelve hours distance from it; and both sending branches into the narrower seas; and the manner and velocity of all these motions will be assigned by means of a map of cotidal lines.

I now proceed to endeavour to determine, first, from the laws of the motion of water, what the form of such lines may be expected to be; second, from the tide observations which we possess, what their form appears to be in reality.

### Sect. I. *On Cotidal Lines as determined by the laws of fluids.*

1. *Tides on a globe covered with water.*—If we suppose the whole surface of the terrestrial globe to be uniformly covered with water, it is easy to see what must be the nature of the form and motion of the cotidal lines. The tides would be, in their main circumstances, entirely governed by the moon. High water at every place, in the same latitude, would follow the transit of the moon at the same interval of time*. The points at which it was high water at a given moment would therefore be situated in a meridian, at a certain distance from the meridian in which the moon was (or at least in some curve symmetrical with regard to the equator). There would be one such curve having reference to the moon, and another, having reference to the point

---

* We here consider the moon as moving in the equinoctial.

immediately opposite to the moon ; and these curves would each revolve round the earth, from east to west, in something more than twenty-four hours. If we suppose one cotidal line to be drawn through all points at which it is high water at 1 o'clock on a given day, a second cotidal line through all places where it is high water at 2 o'clock on the same day, and so on, there will be twenty-four such similar lines on the whole surface of the globe, cutting the equator at equal intervals, like so many meridians. And since the circumference of the earth is about 25,000 miles, it is obvious that any one of these cotidal lines would travel with a velocity of above one thousand miles an hour at the equator, and with a velocity of about six hundred miles an hour in our latitude. This is the velocity with which the summit of the tide-wave would travel on this supposition.

2. *Derivative tides.*—If on such a globe as we have been considering, a continent were interposed, occupying a great extent of latitude, it is clear that the motion of the cotidal lines must become quite different from what it was in the uninterrupted ocean. On the western side of such a continent the tide-wave could no longer proceed as if the continent were not there ; for the supply of water and of pressure brought by the tide-wave advancing from the east, on which its further motion westwards altogether depends, is entirely intercepted. The tide on the western side of the continent must be produced by the water and the pressure which comes from the north, south, and west, and will be governed by laws different from those which regulate the primary or uninterrupted tide. And the same may be said of the tides produced in any seas of which the extent is much intercepted by land.

In order to see the general character of such cases, let us take the case of a tide which is entirely derived from the primary tide, and is not affected at all by the direct action of the sun and moon. Suppose the surface of the southern hemisphere to be entirely occupied by water, and the northern hemisphere to be principally land. Let a considerable inland sea run northward from the equator towards the pole. The tide-wave of the southern ocean, as it passes the entrance of this sea, will send off a derivative undulation, which will advance northwards up the sea, being impelled entirely by the mechanical action by which undulations are propagated in fluids. If we suppose the depth and other circumstances which would affect the motion of this derivative wave to be uniform in different parts of the sea, the wave will advance in the direction in which it first sets out, and therefore, if the entrance of the sea be narrow, in the direction of the length of the sea.   This wave will bring a tide wherever it arrives, and the cotidal lines so produced will be nearly perpendicular to the length of the sea.   The velocity with which the wave moves will depend on various circumstances, but principally on the depth, and probably on the regularity of the channel.   If the depth be nearly uniform, the cotidal lines will be nearly straight and parallel.   Their rectilinear and parallel character depend on this; that the propagation of a wave may be conceived to result from the propagation of undulations in every direction from every point of the line of the wave; and the assemblage of the undulations so propagated, after any interval of time, constitutes the wave in its new position.   Hence, if there be any part of the sea into which undulations are propagated more slowly than they are into the other parts, the wave in that part will not advance so fast as in the rest, and the line of the wave will there hang back.   Thus, if the wave travel more slowly near the shores than in the wide sea, the wave-lines will bend backwards in those parts so as to assume a convex form; and the cotidal lines might resemble the curves I, II, III, IV, V, VI, VII, in the adjoining fig. 1.

Fig. 1.

Fig. 2.

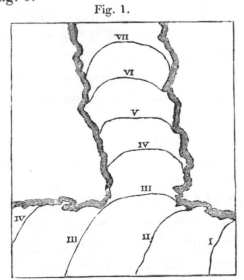
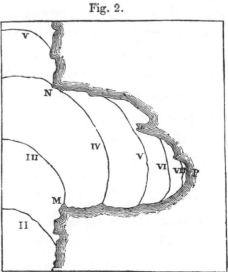

3. *Effect of arms of the sea and bays.*—In the same manner in which the undulation produced in our hypothetical southern ocean sent off a ramification

northwards into the inland sea, the undulation produced in this sea would send off a ramification into any lateral channel or inlet which might branch off from the main expanse.   It will however be proper to consider this case a little further.   Let there be a deep bay on the eastern shore of the inland sea, as fig. 2 ; when the undulation, travelling northwards, reaches the southern cape of this bay, it will be propagated eastwards into the bay as well as northwards, proceeding in all directions from the southern cape (M) till it meets the cape which forms the northern point of the bay (N); after which, the undulation in the main sea and in the bay will be detached from each other, and will each advance independently.   And each of these undulations will again be affected by the form of the shores and other circumstances, in the same way as the main undulation.   It is clear that if we proceed from the point N along the coast in either direction, we shall arrive at points where the tide is later than it is at N ;  the tide-wave separates into two at that point, and N *is a point of divergence of cotidal lines.*

Also, if P be the extreme point which the tide-wave reaches, the tide at P will be later than it is at the coast on either side of P.  The tide-wave travels along the shore to P from each side, and P *is a point of convergence of cotidal lines.*

Also, the velocity with which the undulations advance depends upon the depth of the water, and probably in some measure on the friction and uneven-ness of the sides and bottom.   And as in narrower seas the depth is generally less, and as the effect of the shores will there bear a greater ratio to the whole forces, the velocity in narrow seas and bays will be less than in the main sea. Hence the cotidal lines drawn for equal intervals of time,—as for instance, for intervals of one hour,—will be nearer to each other in narrow seas and bays than in the wider seas.   It will be seen hereafter that in advancing from the southern Atlantic into the German Ocean, the horary intervals of the cotidal lines become less than one twelfth of their original magnitude.

4. *Effect of detached islands and groups of islands.*—The tide-wave, by tra-velling more slowly along the shore than in the open sea, becomes convex forwards.   From this consideration, we can deduce the effect of an island interposed in the space over which it is to pass.   In fig. 3. the I and II o'clock lines are not at all affected, or very slightly, by the island; the III o'clock line is held back so that it meets the shores of the island, although in other

parts of the ocean it is in advance of that place. The same is the case with the IV o'clock line, but the advance being greater, the two convex portions at the two ends of the island are turned towards each other; in the V o'clock hour line these portions touch; and thus the line may be considered as formed of two, which meet at the point of contact just mentioned, one line having its two ends on the shores of the island; the other line running across the ocean like the uninterrupted lines, but with an indentation towards the island. After this time these two lines give rise to two separate waves, 6 and VI; the former moving in a retrograde direction towards the island; the latter moving forwards, and gradually obliterating the indentation produced by the island.

It appears in this way that there is *a point of divergence of cotidal lines* on the side of the island which is towards the coming tide-wave, and *a point of convergence* on the opposite side.

Fig. 3.　　　　　　　　　　　　　　　　Fig. 4.

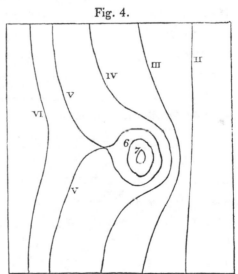

And if there be shallower parts of the ocean, not connected with any land, or connected only with small islands, the effect upon the form of the cotidal lines will be of the same nature, but may go still further. See fig. 4. In advancing upon such a part of the ocean, the cotidal lines immediately behind it will be brought closer together, while to the right and left of the place they will proceed without a corresponding thronging. Hence the cotidal curve on the two sides will advance beyond the islands, while it cannot pass directly over the islands themselves. The undulation will be propagated from the right and left into the space beyond the islands, and the convexities of the cotidal curves will at last meet there, as is shown in the cotidal curves V V in fig. 4. By this means the islands are surrounded by a ring-formed wave, which will advance towards the centre of the ring, and thus produce concentric ring-formed cotidal lines, as 6, 7. In the mean time, as the wave advances beyond the islands after its two parts are united, the indentation in its convexity will be gradually obliterated; and after a time, if the sea be of sufficient extent and of nearly uniform depth, the curve will again become continuously convex.

If the passage on one side of an island be much wider and deeper than on the other, the tide-wave, travelling much more rapidly in the wider space, and always extending itself laterally where there is room, may go round the island on its open side, and return on the other side in a direction opposite to its original one. This is the case with the tide-wave which visits the British Islands, as will be seen hereafter.

5. *Effects of the interference of undulations.*—In such a case as the one just mentioned, the tides going round the island by different paths will at last meet, and the water will be affected by their combined influence. Though we cannot at present refer with certainty to any cases in which the phenomena of the tides in detail are accounted for by considerations of this kind, it will be proper to point out in some measure what the consequences of such a state of things might be.

Let it be supposed that we have a channel into which the tides enter at both ends. Each undulation will be propagated independently of the other; and each portion of the water will be affected by the sum of the elevations and depressions due to both undulations. The undulations proceed in opposite directions, and their velocities, so far as they depend upon the circumstances of the channel, will be equal. If we also suppose their total elevations to be equal, we may trace the result in the following manner:

| I | II | III | IV | V | VI | VII | VIII | IX | X | XI | XII | I | II | III |
|---|----|-----|----|----|----|-----|------|----|----|----|-----|----|----|-----|
| XI | X | IX | VIII | VII | VI | V | IV | III | II | I | XII | XI | X | IX |
| XII | XII | A | VI | VI | VI | VI | VI | B | XII | XII | XII | XII | XII | C |

Let the numerals in the upper line represent the positions of the wave which advances from left to right at those hours; the figures in the next line the positions of the other wave; and the third line the hours of the resulting tide. Two tides arrive at the same point, one having its greatest elevation at XI, and the other at I o'clock, the corresponding low water being at V and VII respectively. But it is clear that from XI to XII, the tide of which the hour is I, will rise faster than the tide of which the hour is XI falls, because the latter is close to its maximum; and thus, by the combined effect of the two undulations, the water will continue to rise; at XII the joint tide will be highest; and in the same way it will appear that the lowest water at this place will occur at VI o'clock. In like manner the two tides which would take place separately at X and at II will produce a joint tide at XII, the intermediate hour, and low water at VI; but the high water will be less high than the former, because both the component tides are further from their maxima. The two tides which would take place at IX and at III might produce a joint tide at XII also, if there were no recurrence of the tides; but the tide which comes at IX returns again twelve (tidal) hours afterwards: the fall of one of the component tides exactly counterbalances the rise of the other; and at this point there is no tide at all, the water remaining always at the same level. Beyond this point, we come to a place where the times of the compound tides are IV and VIII, which would produce high water at the intermediate hour VI; then to a point where the times are V and VII, and the high water (which will be greater than the last) still at VI; then to a place where the two tides coincide at VI, and the tide will be still at that hour, and greater than the adjacent tides; the times VII and V, VIII and IV still give the tide at VI, but successively less and less; the times IX and III give no tide, for the same reason as before. After this point, X and II again give XII for the tide hour, which continues while the component tides are XI and I, XII and XII, I and IX, II and X; till III and IX again obliterate the tide; and after this, it again happens at VI. Thus from A to B the compound tide is at VI, from B to C it is at XII, and so on alternately; and in proceeding from A to B or from B to C the tide increases, attains a maximum amount at the middle point between the extreme points, and then diminishes. In the case of such an interference of two tides, it thus appears that there are not progressive cotidal lines, but stationary cotidal spaces. The portions of water from A to B, and from B to C, rise and fall alternately, while the points A, B, C neither rise nor fall. There are two tide hours only (in this instance VI and XII) over the whole extent of the channel, while some points have no tide at all.

The circumstances just described result from the interference of two tides which move in opposite directions, the two being equal in amount at every point which they reach. If two tides interfere without being equal, they will still produce a compound tide, of which the circumstances may be altogether different from those of a simple tide.

It must be observed, that the propagation of undulations in any direction does not imply necessarily a movement of the water in that direction; therefore the concourse of two undulations coming in opposite directions does not imply the meeting of two currents. In the case where the cotidal lines approach from opposite quarters and come together, we may, if we choose, speak of " the meeting of the tides"; but it is to be recollected, that such a phrase is then used in a different sense from that in which it is often applied; for in common language, the *tides* are said to *meet* at those places where the *current* which brings the flood changes its direction, coming on one side from one quarter, and on the other side from the opposite quarter. How far this *meeting of the tide-currents* may be expected to *coincide* with the *meeting of the cotidal lines*, is a question to be considered hereafter.

We shall endeavour to trace the course of the cotidal lines according to which the tide is actually propagated in the Ocean; and it will be seen that they exemplify most of the preceding remarks. But before we proceed to do this, it will be proper to make a few remarks respecting the tide observations which we possess, and which must be our materials for such an attempt.

Sect. II. *On the causes of inaccuracy in Tide Observations.*

1. *Difference between the time of High Water and the time of Slack Water.*— When we begin to compare the times of the tide at different places as stated by the best authorities, in the usual manner, we find a great number of cases which seem to interfere altogether with any notion of obvious connexion and simple laws prevailing in this class of facts. Thus, if we refer to these authorities as recorded on Mr. LUBBOCK's chart, published in the Philosophical Transactions for 1831, we find many instances where places very near each other are marked with very different hours. The hour at the Eddystone is

marked VIII, while on the adjacent coast of Devonshire it is V ; on the coast at the Land's End it is IV$\frac{1}{2}$, but at a very little distance it is V$\frac{1}{2}$.  On the north coast of Ireland IV$\frac{1}{2}$ and IX$\frac{1}{4}$ stand close to each other ; and similar discrepancies appear in many other places.  Such discrepancies seem at first to make it impossible that the cotidal waves should have any regularity in their form and order.  But there is an additional difficulty with regard to these cases : we can hardly conceive how they are possible on the principle of the water finding its level.  If the time of high water at Plymouth be 5$^h$ and at the Eddystone 8$^h$, the water must be falling for three hours on the shore, while it is rising at the same time at ten or twelve miles distance ; and this through a height of several feet.  We can hardly imagine that any elevation in one of the situations should not be transferred to the other in a much shorter time than this.

There is, in fact, no doubt that most, or all the statements of such discrepancies, are founded in a mistake arising from the comparison of two different phenomena ; namely, the *time of high water*, and the *time of the change from the flow to the ebb current*.  In some cases the one, and in some the other of these times has been observed as the *time of the tide*, and in this manner have arisen such anomalies as have been mentioned.

The time of the change of current, or the *time of slack water*, as it may be termed, *never* coincides with the *time of high water*, except close in upon the shore, and within its influence ; the interval of the two times is generally considerable.  Great confusion has been produced by these two times not being properly distinguished ; so great indeed, that it has made almost all the tide observations which we possess of doubtful value for our present purpose.

The persuasion that, in waters affected by tides, the water rises while it runs one way, and falls while it runs the opposite way, though wholly erroneous, is very general.  Thus it is often stated that the time of tide in the British Channel must be three hours later in the mid-sea than it is on the shore, because the easterly current continues running three hours after the time of high water on the coast.  It would be easy to multiply instances of the perplexities into which various persons have been led by this assumption.  It may suffice to quote one mentioned by Mr. STEVENSON in his account of the Bell Rock Lighthouse.  The waters of the river Dee at Aberdeen, even at the entrance of the harbour, have almost a constant current seaward, notwithstanding the opposite direction of the flood tide of the ocean. One of Mr. Stevenson's assistants, a very intelligent shipmaster, stationed at low water mark, " continued at his post while the water flowed up to his middle ; and when accosted about his situation, he significantly observed, that it was rather extraordinary, as the stream had never ceased to indicate the continuance of the ebb-tide while the water was still rising upon his body."

In this case the current by which the water was affected was not the effect of the tide alone, which generally produces two opposite currents, alternating, and nearly equal in their duration. But in the case of such alternate tide currents also, the time of change of their direction is not in general the time of high water. This indeed is well known to intelligent seamen, who have accordingly phrases by which they express the relation of these times. When the stream in the offing which brought the tide continues to run for three hours after it is high water, it is said to make " tide and half tide *." But though seamen have thus noticed the fact, their recorded observations are very far from always recognising the distinction, as will be seen when we proceed to examine them. In the cases above mentioned, 5 o'clock at Plymouth is the time of high water there ; and 8 o'clock at the Eddystone is the time of the slack water or change of current at that rock ; the time of high water there being, on the average, a few minutes earlier than it is at Plymouth. In like manner $4\frac{1}{2}$ is the time of high water at the Land's End, and $6\frac{1}{2}$ is the time of a certain change in the direction of the current which is assumed to be the change from flow to ebb : in this latter case, however, the assumption is somewhat arbitrary, as the motion of the current is not alternately opposite, the change of its direction being rotatory.

A very little consideration is sufficient to show that in a bay or harbour the time of high water must coincide with the time of slack water, and that in the open sea these times will not coincide. In harbour, the water which flows in at the mouth has no egress by any other part; it therefore accumulates and rises as long as the ingress continues. But in a channel open at both ends, the case is quite different. The rise or fall of the water in any part in such case depends upon this ;—whether the water comes faster than it goes away,

---

* Captain White adds that when the flood stream runs an hour and a half after high water, it is termed " tide and quarter tide ;" and when it runs three quarters of an hour after high water, " tide and half-quarter tide."—Survey, p. 269.

or the reverse. Let the flood in a certain channel come from the west; then at any given point the surface will rise by the water which the flood current brings, and sink by the water which the same current takes away: it will rise therefore if the current, at points to the westward of the given point, be more rapid than it is at points to the eastward of the given point. And the surface will be highest at the given point, when the current is equally rapid at the easterly and at the westerly points; and obviously not when there is no current. And the same is true of the times of lowest water at the given point, considered with reference to the current in the opposite direction. If the channel were of equal width, and the tide of equal height through its whole length, the time of high water would coincide with the time of greatest velocity of the current; and the change of the current would take place at the mean time between high and low water, and consequently six hours after high water. In like manner the time of low water would occur when the current was most rapid in the opposite direction. If the channel be narrower to the eastward, the time of slack water will be later than the time of high water by a smaller interval; and this diminution of the interval goes on till the times coincide, when there is no outlet, as has already been said.

In open seas the time of the change of the direction of the current is of more consequence to the seaman than the time of high water; and accordingly the first has been often recorded, when the other has not: hence the observations, by means of which we are to trace the cotidal lines are more scanty than they at first appear; and, what is worse, are often very doubtful in their meaning. But having pointed out the nature and frequency of this ambiguity, the anomalies which may occur in tide observations will be less perplexing than they were while such a source of confusion was not adverted to; and we may sometimes in such cases be able to distinguish the true from the erroneous statement. The time of high water is the fact which is most important for our present purpose; but that and the time of slack water ought to be both separately noted in all careful observations of the tides.

2. *The change of the Moon's angular distance from the Sun in the course of the day.*—The times spoken of in the preceding paragraphs as recorded for different places are the *hours of the tide on the days of full and new moon*, which times are often called the *establishments* of the places to which they belong.

This establishment is supposed to regulate the time of the tide on all other days of the lunation, the time of the tide being primarily governed by the moon. This supposition, which is true as a first approximation, assumes that the tide always occurs at the same hour-angle from the moon. But the hour of the tide on any day expresses its hour-angle from the sun; and as the moon changes her right ascension by about 48 minutes every day, the observed hour of the tide being given, on the day of full and new moon, the hour-angle from the moon may be different according to the time of the day when the conjunction takes place, compared with the time of day when the observed tide takes place. Thus if the conjunction take place at 1 o'clock in the morning, and the observed tide at $11^h$ at night, the distance of the tide from the sun is eleven hours; but at $11^h$ at night the moon is to the east of the sun by her motion in 22 hours, which is 44 minutes of hour-angle, and therefore the tide is only 10 hours 16 minutes behind the moon. But if the observed tide take place at $1^h$ in the morning and the conjunction at $11^h$ at night, the moon, at the time of the tide, is 44 minutes to the west of the sun, and the tide is 1 hour 44 minutes. In the former case the establishment is 44 minutes less, in the latter 44 minutes more, than the observation of the *hour* of the tide gives it.

If the time of tide were observed to be $6^h$ in the evening, the conjunction being at $1^h$ in the morning, the true establishment is $5^h$ $26^m$, but if the tide be at $6^h$ in the morning and the conjunction at $11^h$ in the evening, the true establishment is $6^h$ $34^m$. In this way it appears that an observation of the hour of the tide on the day of new or full moon leaves an uncertainty of 1 hour 8 minutes (and it may be more) as to the establishment, if we do not take into account whether the morning or afternoon tide was observed, and at what hour the conjunction or opposition of the moon took place.

In addition to this ground of uncertainty, the time of high water may often be doubtful to the extent of ten minutes or a quarter of an hour, from the want of precision in the observation; and as this error may occur in opposite directions at two different observations, and may be combined with the variation just mentioned, we may have thus two establishments different by above an hour and a half, collected from observations of the same place.

We cannot obtain any considerable accuracy in the determination of the establishment, without using numerous observations; and in this case the mean of the morning and evening tide-hours may be taken, the effect of the intervals by which the conjunctions and oppositions of the moon precede and succeed noon being supposed to counterbalance each other. In this case also the errors of observation may be supposed to be corrected in the average. But if we have to collect the establishment from a few observations, it will be proper to calculate in each case the hour angle by which the tide is distant from the moon.

3. *The semimenstrual inequality of the establishment.*—It has been already said, that the supposition that the tide depends on the moon alone, is a first approximation only. The time of high water does not follow the moon's transit by the same interval at every period of the lunation; the interval is sometimes greater and sometimes less than that *corresponding* to the new and full moon, and is regulated by the distance of the moon from the sun. The following is the mean state of this variation. When the moon and sun are in conjunction, the corresponding tide follows the moon by its mean interval. When the moon is at various hour angles after the sun, the following are the mean corrections of the mean interval, negative and positive*.

Hour angle of the moon,          0   1   2   3   4   5  6  7  8  9  10 11 12 hours.
Correction of the establishment, 0 −16 −31 −41 −44 −31 0 31 44 41 31 16  0 minutes.

Thus, if the establishment corresponding to the new and full moon be 6 hours, the time of the corresponding high water when the moon is 1 hour from the sun, will be $5^h 44^m$ after the moon's transit; when the moon is 2 hours from the sun, the time of tide will be $5^h 29^m$ after the transit; and so on. When the moon is 6 hours from the sun, the corresponding time of high water will again coincide with the mean, after which the interval of the transit and tide will be greater than the mean, till the next conjunction or opposition; and then the same cycle recurs.

Hence, if the establishment were collected from any observation of the tide not corresponding to the day of new or full moon, it would be liable to an error. If the establishment were 6 hours, by an observation made when the moon's hour angle was 4 hours, and compared with the time of the moon's

---

* The law and magnitude of these numbers depend on the relative effect of the sun and moon upon the tides; the amount varies with the declination of the sun and moon, and with the moon's parallax.

MDCCCXXXIII.                              Y

transit, it would appear to be $5^h 16^m$; but by an observation made when the moon's hour angle was 8 hours, it would appear to be $6^h 44^m$.

This cause of difference in the results would be avoided by making the observation of the tide corresponding to new or full moon, or by applying the proper correction, according to the preceding Table, when any other tide was observed. The chance of error would however be removed more effectually by taking the mean of all the intervals between tide and transit, corresponding to half a lunation, or to any whole number of half-lunations.

4. *The correction of the establishment for the age of the tide.*—In the preceding paragraph we have spoken of the tide *corresponding* to new or full moon, and not of the tide which takes place *on the day of* new or full moon. The latter however is that which has been commonly observed for the purpose of determining the establishment of any place; but it does not coincide with the former, and there are certain anomalies in the tide records, depending on this difference.

The tide which comes to the shores of narrow and long seas is not immediately produced by the moon, but is derived from the tide in the main ocean; its circumstances are governed by those of the primary tide from which it is derived, and whatever interval may be employed in its transfer, it is regulated by the position which the sun and moon had at the time when they determined the primary tide. Now, this time may have been one, or two, or more days, before the tide reaches the place where it is observed. Thus, the tide on the shores of North America and Spain is determined by the configuration of the sun and moon at a day and a half previous; the tide in the port of London appears to be two days and a half old when it arrives. This circumstance affects the determination of the establishment from observations, in a manner which must be explained.

Since the tide at London is determined by the position of the sun and moon $2\frac{1}{2}$ days before it occurs, the moon must then have been more to the west of the sun by an hour angle of 2 hours (her motion in Æ. in $2\frac{1}{2}$ days,) than she is when the tide arrives. Hence, the tide which happens on the day of full moon corresponds to the period when the moon was in Æ. 2 hours west of the point opposite to the sun, or 10 hours east of the sun. Therefore, by the Table in last page, the tide is 31 minutes later than the mean interval of tide and moon's transit.    The tide is observed to take place at 2 o'clock on the days of new and full moon, therefore $1^h 29^m$ is the *corrected establishment* for London.

In general, however, the establishment is defined to mean the hour of high water *at* new and full moon.    We shall call this the *vulgar establishment*. Observations of tides have generally been directed to the object of determining this vulgar establishment, which, it appears, by what has been said, is not a corresponding quantity at different places.    The mean of all the intervals of tide and transit for a half-lunation is the corrected establishment; the vulgar establishment is greater than this by a quantity depending on what may be called *the age of the tide,* namely, the length of time which has elapsed since its real or theoretical origin.

The corrected establishment may be determined, as has already been said, by taking the mean of the intervals of tide and moon's transit for any whole number of half-lunations.    But we may observe, that it might be collected immediately from the vulgar establishment, if we had obtained a first approximation to the distribution of cotidal lines upon the surface of the ocean; for the age of the original tide in any part of the open ocean being known, the age of the tide derived from the original tide in any other part would be known from the number of intervening cotidal lines.    Thus, if the tide on the coast of Spain be a day and a half old, the tide on the coast of Norfolk must be nearly two days and a half old, inasmuch as there are nearly 24 horary cotidal lines in the interval, if we follow the sea round the north point of Scotland, which is the course by which the tide reaches the eastern coast of England.

Since the recorded tide observations are liable to the very great inaccuracies and even ambiguities which have been pointed out, it may easily be conceived that we cannot at present deduce from them the course of the cotidal lines with accuracy and certainty.    We may add to this, that our observations are very scanty in extent, compared with those which such a use of them would require; there are many seas and coasts where we have no information at all respecting the times of tide.    As, however, an attempt to use such observations as we have, may, perhaps, lead to the collection of more numerous and more accurate ones, I shall endeavour to draw a first approximation to the course of the cotidal lines, begging the reader to bear in mind that from the nature of our materials it must be imperfect, and may be widely erroneous.

Sect. III. *Discussion of the Tide Observations now extant.*

There are various sources and collections of information on the subject of tide observations. The most important essay towards a complete collection is that contained in the 4th volume of LALANDE's Astronomy. In this the author has not adverted to the causes of confusion which have been pointed out; and his survey of the existing information at that period led him to terminate his statement with an earnest request that all persons having the opportunity would endeavour to render our knowledge more complete. In various books of astronomy and navigation, there are lists of the establishments of places in different parts of the world; and in Sailing Directions for considerable tracts of the ocean, statements of the same kind are collected. Among works of this nature I may mention NORIE's Epitome of Navigation, and PURDY's three Memoirs, that on the Atlantic Ocean, that on the Ethiopic or South Atlantic, and his Columbian Navigator. The Sailing Directions for more limited spaces also contain such statements. The nautical surveys of various navigators supply the establishments of places recorded in the charts, or in the accompanying remarks; I may notice particularly the Surveys of the Australian coasts by Captains FLINDERS and KING, and of Patagonia by the latter officer, in addition to many others nearer home. The "Remark Books" of various ships contain many such observations. These latter documents, and a large proportion of the information contained in the various surveys which have been almost unintermittingly carried on by naval officers of this country for a long course of years, exist in manuscript in the Admiralty.

With regard to these materials—I have, by the kindness of the Hydrographer, Capt. BEAUFORT, been allowed the free use of the charts and manuscripts belonging to his department, without which advantage, indeed, I should hardly have been able to make the present attempt, imperfect as it may be.

I shall begin by considering the tides of the Atlantic, which are, at least in their main features, of a derivative kind, and are propagated from south to north according to the laws of undulations in a limited sea, as explained in Sect. I. of this memoir.

The cotidal lines, which I shall draw by means of the data I am now about to discuss, are drawn through or near the points for which the time of high water is supposed to be ascertained; and are moreover drawn so as to possess as much regularity and similarity in their form and intervals, as the data will allow. By this means the lines pass across parts of the ocean where no tide observations have been, or perhaps can ever be made. The tide hours in such parts are to be considered as obtained by the interpolation to which the forms of the cotidal curves conduct us.

### The East Coast of the Atlantic.

If we look at Mr. LUBBOCK's chart of the world, or at any other good general view of the tides along the west coast of Africa, Spain, Ireland and Scotland, it will appear tolerably certain, from the tide-hours given for different places, that the tide-hour which is about $1\frac{1}{2}^h$, Greenwich time, at the Cape, becomes successively, in going northwards, $2^h$, $3^h$, $4^h$, $5^h$, $6^h$, $7^h$, $8^h$, $9^h$, $10^h$, $11^h$, $12^h$ which is the hour about Cape Blanco, and goes on to $2^h$, $3^h$, $4^h$, $5^h$, &c. on the western coasts of Europe.

For the greater part of this coast we have not the means of determining the hour with much additional accuracy. Concerning the Cape of Good Hope, where we might have expected the establishment to be well and accurately known, I have not been able to obtain good information. Certain " Observations " sent from that place to the Admiralty are obviously not to be depended on. The " Tide Table calculated for Table Bay " inserted in the South African Almanac for 1832, must be very erroneous for a large proportion of days, even if the mean establishment be right. This will be clear when it is stated that the Table is constructed by assuming certain tide-hours for each day of a cycle of 30 days, and by applying this cycle, with no exact regard to the day of the moon's age, and with no regard at all to the hour of her transit. In NORIE's Epitome of Navigation, the establishment of Table Bay is given as $2^h 25^m$, and subtracting $1^h 14^m$ of east longitude, we have $1^h 11^m$ for the establishment reduced to Greenwich time. I shall for the present adopt this value.

The statements concerning the tides at St. Helena are various. NORIE's Epitome gives the time for James Town, as $1^h 30^m$. Dr. MASKELYNE in 1761 made a series of observations on the tides during the months of November and December, (Phil. Trans. 1762, p. 586.). These are very irregular, which is the less surprising as the observation was made by means of a post fixed in a part of the harbour where the waves were a foot or two high.  I have compared these observations with the times of the moon's southing in the *Connaissance des Tems* for 1761, and find that they give a mean establishment of about $2^h$.

Certain observations made at this island in September 1826 by Lieut. JOHNSON have been communicated to me; they are very irregular, probably in consequence partly of uncertainty arising from the smallness of the tide, the greatest rise not being much more than three feet. The establishment resulting from these appears to be about $2^h$.

There are, however, among the Admiralty papers some observations made by General WALKER at the request of Mr. FALLOWS, which from the precautions with which they are described to have been made, and from the accordance of the results, appear to be more worthy of confidence than any of those previously mentioned.  These have been examined by Mr. DESSIOU, and have given as the mean of the semilunations, fourteen in number, $2^h 55^m$ for the hour by which the tide follows the transit of the moon. I shall therefore adopt this as the establishment at St. Helena.  The difference of longitude from Greenwich is so small that no correction on that account is requisite.

The tide hour at Ascension Island is stated in two "Remark Books" of different ships, as $4^h$ and $5^h 30^m$.  I have been furnished with a record of observations made by Capt. R. CAMPBELL, R.N., at various periods from March 1820 to August 1821, by which it appears that the establishment is about $5^h 5^m$; adding to this $57^m$ of west longitude, we find that the cotidal line of $6^h 2^m$ touches this island.

Returning to the coast of Africa, and taking for our authority Mr. LUBBOCK's chart, which in this part he states to be founded on Captain OWEN's surveys, and referring also to the list in NORIE's Epitome, we find that we have the following order of latitudes and establishments.

|  | Latitude. | Establishment. |
|---|---|---|
| Saldanha Bay | 33° 2′ S. | $2^h 0^m$ |
| St. Helena Bay | 32 42 S. | 2 30 |
| Cape Serra | 22 0 S. | 3 0 |
| St. Paul de Loando | 8 48 S. | 4 30 |
| Gaboon River | 0 30 N. | 5 0 |
| New Calebar River | 4 22 N. | 5 0 |

The coast during this space runs nearly north and south, so that the correction for longitude will not much affect the differences of these times; and hence it appears tolerably certain that the tide-wave travels from the Cape of Good Hope to the bottom of the Gulf of Guinea in something less than four hours.

We have the following statements for points in the Gold Coast.

|  | Longitude. | Establishment. |
|---|---|---|
| Cape Coast Castle | 4′ E. | $3^h$ $30^m$ NORIE. |
| Cape Three Points | 9 W. | 3 30 LUBBOCK. |

This would seem to imply that the $3^h$ $30^m$ cotidal line crosses from the neighbourhood of St. Paul de Loando to the Gold Coast, which would not be impossible, but appears to be inconsistent with the tide hours at St. Helena and Ascension: I shall therefore not draw it so till we have more certain information.

The following tide hours are given for places in Fernando Po by NORIE.

| George's Bay | $4^h$ $0^m$ |
|---|---|
| Goat Island | 4 0 |
| Cape Buller | 4 15 |
| Also Island of St. Thomas | 3 25 |
| But | 5 30 in Mr. LUBBOCK's Chart. |

The last-mentioned observation falls in most easily with the general form of the lines. I shall for the present suppose it to be correct, and the others inaccurate.

In proceeding westward along the coast I do not know any statements of the tide hour till we come to Sherbro Island, the Islands of Bananas, Cape Sierra Leone, and the Islands of Los. These places lie between latitude 7° 35′ and 9° 30′ N., and have west longitude about $52^m$ of time. Their establishments are stated as follows:

Sherbro Island . $5^h$ $53^m$; and reduced to Greenwich time, $6^h$ $45^m$ LUBBOCK.
Bananas Islands . 8 15 PURDY *.
Sherbro River . . 8 0 NORIE.

* Memoir to accompany a new Chart of the Atlantic Ocean, 1820.

| | Establishment. | | Greenwich Time. |
|---|---|---|---|
| Cape Sierra Leone . . | 7$^h$ 30$^m$ NORIE . . . | 8$^h$ 22$^m$, but 7$^h$ 45$^m$ LUBBOCK. |
| River Sierra Leone . | 8 15 PURDY. | |
| Isles of Los . . . . | 9 0 —— . . . | 9 56, but 7 30 LUBBOCK. |
| | 7 40 BOTELER*. . | 8 32 |

Taking Captain BOTELER's times as more probable in this case, we perceive that the 7$^h$ tide line must fall somewhere in the neighbourhood of Sierra Leone. Thus, if we had adopted the hour 3½$^h$ stated for Cape Coast Castle, it would follow that the tide-wave occupies 3½ hours in moving from that place to Cape Sierra Leone,—an interval of less than one hour of longitude. On the other side of the Gulf of Guinea the tide-wave in the same time, namely, from 1$^h$ to 4½$^h$, had moved through about forty degrees of latitude.

As we proceed to the north the times become later, though without any apparent regularity. We have the following hours given :

| | Latitude. | Establishment. | |
|---|---|---|---|
| Bathurst . . . . | 13° 28' N. | 8$^h$ 10$^m$ NORIE. | BOTELER. |
| River Gambia (within) | 13 39 | 11 45 | —— |
| Goree . . . . . | 14 40 | 7 48 | —— |
| | | 7 0 | ADANSON (Savans Etrangers, ii.605.) |
| Cape Verd . . . . | 14 43 | 7 45 | NORIE. |
| Senegal (bar) . . . | 16 1 | 10 30 | —— |
| Cape Blanco . . . | 20 50 | 9 45 | —— |
| River Ouro . . . | 23 51 | 12 0 | —— |
| Cape Bojador . . . | 26 7 | 12 0 | —— |

The observations made within the rivers Senegal and Gambia are apparently affected by the retardation due to the inlet of the river, and may be rejected; that at Bathurst is probably affected in the same way; perhaps that of river Ouro. The longitude of all these places is about 1$^h$ west. Hence I shall take for the establishment of

| Cape Verd . . . . | 8$^h$ 45$^m$ Greenwich time. |
|---|---|
| Cape Blanco . . . | 10 45 |
| Cape Bojador . . . | 1 0 |

* Sailing Directions for the West Coast of Africa.

If these data can be depended upon, the 12$^h$ tidal line meets the coast of Africa somewhere very near latitude 23$\frac{1}{2}$°.

The anomalies of the recorded tide hours of the coast of Africa to the north of this are so perplexing, that I shall in the first place proceed to the coast of Spain, where the order is clearer.

The tide at Cape St. Vincent is stated to occur at 2$^h$ 15$^m$; at Cape Ortegal, at 3$^h$; and at nearly the same time on the south coast of the Bay of Biscay. Opposite Brest the establishment is 3$^h$ 30$^m$, which is also stated to be the tide hour of Valentia at the S.W. corner of Ireland. Beyond this line of 3$^h$ 30$^m$ we are able to trace the course of the tidal wave in great detail on the shores of this and the neighbouring countries; but we shall first endeavour to complete our view of the Atlantic.

*West Coast of the Atlantic.*—I shall begin the examination of the tides on the western shores of the Atlantic Ocean, from Cape Frio, in lat. 22° 59′ S. We possess observations of the tide at this point, which are probably pretty exact, having been made during a stay of considerable length, by the persons engaged in the operations carried on for the purpose of recovering the lading of the treasure-ship Thetis, which was sunk there. In a chart of the port of Cape Frio, by Lieut. H. KELLET, which appeared in the Nautical Magazine for April 1832, the establishment is stated to be 1$^h$ 40$^m$; and as the longitude is 2$^h$ 48$^m$ W., the tide hour, Greenwich time, is 4$^h$ 28$^m$.

M. ROUSSIN, who surveyed this coast in 1819–1820, also gives the hour at the Bay d'Espirito Santo, in lat. 20° 18′ S. as 3$^h$; and at the Island of St. Sebastian, to the south of Cape Frio, in lat. 23° 50′, as 2$^h$; but at Santa Catharina and Rio Janeiro as 2$^h$ 45$^m$ (lat. 27° 30′); so that from St. Sebastian the hour appears to be later both to the north and south, and there is a point of divergence in that neighbourhood: on all accounts Cape Frio is its most probable position.

PURDY (Ethiopic Memoir, p. 59,) gives the establishment at Cape Frio as 9$^h$; and the person who examined that shore after the loss of the Thetis (Admiralty MSS.) states the time of tide as 9$^h$; but it is clear from the context, that he means the time of change in the direction of the stream; which naturally attracted the greatest share of his attention, the shipwreck having been occasioned by the effect of currents.

Proceeding from Cape Frio northwards along the coast of Brazil, I find the following statements :

| | Lat. | H. W. | H. W., Gr. T. |
|---|---|---|---|
| | ° | h m | h m |
| Bahia or St. Salvador .... | 13　0 S. | 4 15 NORIE, ROUSSIN.*.. | 6 49 |
| | | | 6　0 LUBBOCK. |
| Pernambuco............ | 8　4 | 7 15 NORIE............ | 9 35 |
| | | | 7　0 LUBBOCK. |
| Paraiba ................ | ........ | 4 15................. | 6 35 ROUSSIN. |
| Cape St. Roque ........ | 5 28 | | |
| Fernando Noronha ...... | 3 .56 | 4　0 NORIE............ | 6 15 LUBBOCK. |
| Ciara, or Seara.......... | 3 45 | 4 40................. | 7 14 ROUSSIN. |

The establishment given by M. ROUSSIN for Pernambuco appears much more probable on all accounts than NORIE's, and I shall adopt it.

Beyond Cape St. Roque the coast trends away to the westward. The following times of tide are given :

| | Long. | H. W. | H. W. Gr. T. |
|---|---|---|---|
| | h m | h m | h m |
| Cape St. Roque ........ | 2 21 W. | | |
| Jaguarybe.............. | 2 31 | 6　0 ................ | 8 31 Sailing Directions. |
| *Maranham*.............. | 2 56 | 7　0 NORIE, Brazil..... | 9 56 Also ROUSSIN. |
| *Para*................... | 3 14 | { 12　0 NORIE, Epit. | |
| (Mouth of river Amazon). | | { 4　0 NORIE, Brazil. | |
| Cayenne .............. | 3 29 | 4 30 NORIE.......... | 7 59 |
| | | 3 45 PURDY. | |
| Surinam. Bram's Point .. | 3 41 | 5 30 PURDY........... | 9 11 |
| N. Amsterdam.......... | 3 50 | 4 30 ................ | 8 15 LUBBOCK. |
| (River Berbice) | | | |
| Demerary River ........ | 3 52 | 4 30 NORIE........... | 8 22 |
| Barbadoes ............. | 3 59 | scarcely perceptible (PURDY). | |
| Trinidad ............... | ........ | 4 30 PURDY........... | 8 30 LUBBOCK. |
| —— Port Spain ...... | 4　6 | 6 30 ................ | 10 36 NORIE. |
| St. Lucia.............. | 4　4 | ...................... | 10 45 LUBBOCK. |
| Guadaloupe & Martinique. | 4　7 | | |
| (Irregular) .......... | ........ | 6 45 PURDY........... | 10 52 |

The tide at Para is perhaps retarded several hours by the inlet in which it stands, and that at Maranham probably by two or three hours. Supposing this to be the case, we see that the tide-wave advances to the westward and northward with tolerable regularity.

This agrees with RICHER's observation (Acad. Par. vii. Part II. p. 320.), that the tides which are at $3^h 45^m$ at Cayenne, are earlier in proportion as we approach the equator.

* Le Pilote du Brazil, 1819–1820.

Our materials are probably insufficient at present to enable us to make out with any degree of accuracy the course of the tide-wave among the Bahama Isles from Porto Rico to Florida.

I shall, however, insert the following statements, which I find in NORIE, annexing the rise of the water at spring tides, when it is given :

| | H. W. | | |
|---|---|---|---|
| | h | m | Ht. in Feet. |
| Caribbees. | | | |
|   Saintes . . . . . . . . . | 6 | 45 | |
| Porto Rico. | | | |
|   St. Juan . . . . . . . . . | 8 | 20 . . . . | 1½ |
| Hayti. | | | |
|   Cape Haytien . . . . . . . | 6 | 0 . . . . | 2½ |
|   Puerto da Plata . . . . . | 7 | 30 . . . . | 3 |
| Miraporvos . . . . . . . . | 9 | 30 . . . . | 2 |
| Lucayos. | | | |
|   *St. Salvador* . . . . . . | 3 | 50 | |
|   Providence Island . . . . | 7 | 30 | |
|   Bury Island . . . . . . | 7 | 30 | |
| Bahamas. | | | |
|   Exuma Bar . . . . . . | 6 | 35 | |
|   Royal Island Harbour . . . | 7 | 45 . . . . | 3½ |
|   Pelican Harbour . . . . . | 7 | 30 . . . . | 4 |
|   Man of War Kay . . . . . | 8 | 10 . . . . | 4 |

I shall now endeavour to trace the progress of the tide-wave to and along the coast of North America.

The Bermuda Isles are placed in a position where an exact observation of the tides would throw light upon the course of the cotidal lines. Mr. LUBBOCK states the time of tide (Gr. T.) as $11^h 15^m$, so that the 11 o'clock cotidal line must pass to the east of them. I have inspected, at the Admiralty, statements of observations made at the Naval Yard in Bermuda, in August and September 1832, from which it appears, that the time of high water at full and change is

z 2

$7^{\mathrm{h}}$ $18^{\mathrm{m}}$ on the average, which added to $4^{\mathrm{h}}$ $19^{\mathrm{m}}$, the longitude of the place, gives $11^{\mathrm{h}}$ $37^{\mathrm{m}}$ for the Greenwich time of tide. This discrepancy is slight.

For the coast of North America we have, besides other materials, the Tide Tables in the American Almanac. Taking the Almanac for 1831, and comparing the times there given with the moon's southing, we have,

| | Lat. | Long. | H. W. after Moon's southing. | H. W., Gr. T. |
|---|---|---|---|---|
| | ° ′ | h m | h m | h m |
| Charleston.... | 32 45 | 5 20 | 7 22 | 0 42 |
| New York.... | 40 40 | 4 56 | 9 0 | 1 56 |
| Boston ...... | 42 20 | 4 44 | 11 38 | 4 22 |

Hence it appears, that along this coast the tide-wave runs steadily northwards; and this continues all the way to the mouth of the Bay of Fundy, where the tide from the north again makes its appearance, as is seen by considering the tides of Nova Scotia.

By observations of the tides made at Halifax Yard on the east side of Nova Scotia, and transmitted to the Admiralty, it appears that the establishment at that place is about $7^{\mathrm{h}}$ $42^{\mathrm{m}}$. Proceeding southward along this coast, we come to the southern point of Nova Scotia, Cape Sable, where the tide hour is stated to be $8^{\mathrm{h}}$. (Purdy, Atl. Mem. p. 75.) Turning round this Cape we enter the Bay of Fundy, and just within it is Cape St. Mary, where the tide hour is $9^{\mathrm{h}}$: at the entrance of the Gulf of Anapolis, a little further up, the hour is $10^{\mathrm{h}}$; and in advancing further up to the head of the bay it becomes $11^{\mathrm{h}}$ and $12^{\mathrm{h}}$.

On the opposite or western side of this inlet, at Penobscot and the neighbourhood, the establishment is $10^{\mathrm{h}}$ $45^{\mathrm{m}}$; and as at Boston, further south, it is $11^{\mathrm{h}}$ $30^{\mathrm{m}}$ according to Purdy, or $11^{\mathrm{h}}$ $38^{\mathrm{m}}$ according to the Almanac, it appears that near Penobscot there must be *a point of divergence of cotidal lines*, the wave on the right hand running into the Bay of Fundy, and on the left into the Bay of Massachusetts formed by Cape Cod.

It will easily be conceived that the tide lines break into Chesapeak and Delaware Bays as separate inlets :—to follow these subdivisions of the lines would not suit either our present limits or materials.

The tides in the Bay of Fundy are very high, perhaps the highest in the world. In some places the spring tides rise sixty or seventy feet perpendicular. This is accounted for in some measure by the course of the cotidal lines. The tide which is at $7^h$ $42^m$ at Halifax, long. $4^h$ $14^m$ W., is on the cotidal line of $11^h$ $56^m$; and this line again meets the coast of North America somewhere south of Charleston. The whole of the wave which advances from this line is made to converge by the shore of Nova Scotia on one side, and that of the United States on the other, into the entrance of Fundy Bay, and is thus accumulated to an unusual height.

It will be seen, when we come to speak of the tides of South America, that the tide lines on the coast of North America, which appear to be clearly established, agree with those on the coast of Brazil and Patagonia in several of those circumstances which in the last-mentioned case might appear most doubtful. In both instances the tail of the tide-wave, dragging along the shore, carries a large tide to the northward; and in the mean time the convex front of the wave, advancing more rapidly in the open sea, catches a more northern point of the coast, from which it sends another feebler tide southwards to meet the former.

We may add to the preceding data, that the $12^h$ cotidal line, which, as we have seen, passes near the coast of Nova Scotia, also passes near the coast of Newfoundland. At St. John's the establishment is $7^h$ $50^m$ (NORIE), which with the longitude $3^h$ $30^m$, gives $11^h$ $20^m$ for the Greenwich time. Mr. LUBBOCK gives $11^h$ $30^m$. In Placentia Bay, on the south side of the island, PURDY gives $9^h$ $15^m$ for the time, which agrees with Mr. LUBBOCK's $12^h$ $45^m$ Greenwich time.

I shall attempt further to pursue the tide-wave in its course up the gulf and river of St. Lawrence, in speaking of river tides.

We appear now, from what has been said, to have the means of drawing the cotidal lines of the Atlantic in a general way: but before we proceed to do this, there are some anomalies pertaining to the islands which require to be noticed.

### Tides of the Atlantic Islands.

It is evident from what has been already said, that $12^h$ cotidal line runs obliquely across the Atlantic from the neighbourhood of Newfoundland, so as to meet the shore of Africa in about latitude 23° 30′ N. It appears also from the tides of the western shores of Europe, that the $2^h$ and $3^h$ cotidal lines advance in lines nearly parallel to this. But in the intermediate space lie the Azores, the

Madeira and the Canary Islands; and from what was said in Sect I., we should expect that the cotidal lines would be for a space thrown back and inflected in the neighbourhood of these islands, so that the tides will be later than they would be in that part of the ocean if it were deep water. It has also appeared that there may be detached spaces within which the tides are later than in the surrounding seas, occupied by *converging rings or loops of cotidal lines.*

It appears from the materials which we possess, that some of these circumstances do occur with regard to the islands of the Atlantic. The hours given for the Cape Verd Islands are, St. Jago 6$^h$, St. Nicholas 7$^h$, English Road, Bonavista, 7$^h$ 30$^m$; (Norie's Brazil, and Purdy.) If we consider the differences of these as too doubtful to be relied on, we may take 7$^h$ as the establishment for this group, which gives 8$^h$ 30$^m$ for the Greenwich time; and it will be found that the 8$^h$ 30$^m$ cotidal line will pass without much flexure through this part of the ocean. The differences of time at different points of the group will modify these lines in detail.

The hours given for the Azores are, Fayal Road 11$^h$ 30$^m$, Terceira 11$^h$ 45$^m$, (Norie); to which Mr. Lubbock adds, St. Michael 12$^h$ 30$^m$. Probably the former times belong more nearly to the general sea, and the latter (which is to the east of the others,) is perhaps affected by the retardation of the lines in a converging ring or loop. The time 11$^h$ 30$^m$ gives 1$^h$ 30$^m$ Greenwich time for the cotidal line. It will be seen by inspection that this line must here have been thrown behind its general course.

It is stated by Norie and Purdy, that the hour throughout the Canary Islands is about 3$^h$. This shows a considerable retardation, and makes it necessary to suppose here, that the cotidal lines form a converging ring or loop. It is also stated by the same authorities, that the tide hour at Cape Geer on the coast of Africa, nearly opposite the Canaries, is 2$^h$ 15$^m$, and at Mogadore 4$^h$, which confirms the supposition of such a flexure.

At the same time, it would seem that this cotidal loop or ring is not very extensive; for at Madeira, which is only a few degrees north of Canary and a few degrees west of Cape Geer, the hours given are 12$^h$ 4$^m$ and 12$^h$ 15$^m$; and Cape Geer and Mogadore are so near each other, that it is hardly possible that the hours just mentioned for these places should be both right. It is probable, that somewhere between the Strait of Gibraltar and Cape Blanco there is a point of convergence, and that the $2^h$, $3^h$, and $4^h$ cotidal lines form loops among the Canaries, but we cannot at present pretend to determine their course more accurately.

The tide time at Cape Vincent ($2^h$ $15^m$) places the cotidal line of $2^h$ $51^m$ there, and the tides here must come principally from the west; the southern supply being much interrupted by the islands, as we have seen. And hence the tide must run into the gut of Gibraltar from the west, and we should expect to find the times later and later in approaching the Strait. The statements generally given are, Bay of Cadiz $1^h$ $45^m$*, Portal of Cadiz $2^h$ $15^m$, Zanta Island $12^h$, Gibraltar $12^h$ $15^m$. These hours appear to imply some unexplained anomaly or inaccuracy. It is scarcely possible that the cotidal line of $12^h$ $30^m$, or even of $1^h$, should have a flexure which carries it to Gibraltar. NORIE gives for Cadiz $2^h$ $30^m$ (marking it *Obs.*), which is reconcilable with the general course of the lines, and for Cape Spartel, on the south of the Strait, he gives $3^h$.

After approaching the coast of Spain, the cotidal lines appear to have a form in which the indentation produced by the islands is obliterated, as we should expect from the principles above laid down.

### Tides of the North Sea.

We have already seen that the tide approaches the shores of our own and neighbouring countries from the south-west †; we have now to trace its course more particularly after it reaches the land.

The cotidal line of $4^h$, Greenwich time, of which the direction is about N.W. and S.E., appears to be nearly that which first touches the coasts of Britanny and Ireland. We have,

---

* LALANDE, p. 321, gives for Cadiz $1^h$ $10^m$ " according to twenty-four observations which I received from Cadiz made in 1773 by M. TOFINO."

† An opinion formerly prevailed that the tides came to the coasts of England and Ireland from the north. Mr. MURDOCK MACKENZIE, in his Maritime Survey of Ireland, shows that the flood comes from the Atlantic to the west, divides itself into three streams at the S.W. points of Ireland and England, of which streams two meet again at the N.E. point of Ireland. SPENCE's Scilly Isles, p. 6, (1792).

On the coast of France,

|           | Time, H. W. | Long. W. | Gr. Time. |                                      |
|-----------|-------------|----------|-----------|--------------------------------------|
|           | h   m       | m        | h   m     |                                      |
| Ushant......... | 3   47 | 20     | 4   7     | DAUSSY, (Conn. des Tems, 1834, p. 75). |
| Brest ......... | 3   48 | 18     | 4   6     | ———— Ibid.                          |

And on the S.W. coast of Ireland,

|              | h   m | m   | h   m  |              |
|--------------|-------|-----|--------|--------------|
| Skellings ...... | 3   30 | 40 | 4   10 | SPENCE, p. 8. |
| Valentia ...... | 3   30 | 40 | 4   10 | LUBBOCK.     |
| Cape Clear .... | 4   0  | 38 | 4   38 | MACKENZIE.   |
|              |       |     | 4   45 | LUBBOCK.     |

In the last instance we already see the effect of the retardation produced in entering St. George's Channel.

We shall trace the tide along both this and the British Channel; but we may notice in the first place the Bay of Biscay. As materials for the lines on this coast, we possess the results of a " Reconnaissance Hydrographique des Côtes de France," undertaken by the corps of Ingénieurs-hydrographes, under the direction of M. BEAUTEMS BEAUPRÉ; of which the results, so far as the tides are concerned, have been stated by M. DAUSSY in a memoir inserted in the Connaissance des Tems for 1834. It might have been expected that the tide would take some time to reach the head of this bay at St. Jean de Leon. It appears, however, that the tide is nearly contemporaneous along the whole coast of the bay. The following times are given by M. DAUSSY, among many others:

| | h   m |
|---|---|
| Isle of Noirmoutier (Mouth of the Loire) . . . . | 3   15 |
| St. Martin-de-Ré (Isle of Ré) . . . . . | 3   40 |
| La Rochelle . . . . . . . . . . | 3   39 |
| Tour de Cordovan (Mouth of the Gironde) . | 3   59 |
| Socoa (head of the Bay) . . . . . . . | 3   31 |

The $3^h\ 30^m$ cotidal line must therefore be nearly parallel to the coast of France.

### Tides of the British Channel.

From this point there is no difficulty in tracing the course of the tide-wave along the British Channel, if in doing so we avoid the confusion already noticed, which is often made between the time of high water and the time of change from ebb to flow current; and if we also leave out of our account, or consider with proper allowances, tides in shores very much inbent, as the Bay of Poole and the Solent Sea. Thus, to determine the cotidal lines which pass across the Channel, we must not take the tides at Portsmouth, but those on the outside of the Isle of Wight.

Proceeding to the Scilly Isles, we find the time noted $4^h 10^m$, which gives Greenwich time $4^h 35^m$; Mr. Lubbock has $4^h 30^m$.

At the Land's End the time is $4^h 20^m$, which gives Greenwich time $4^h 43^m$. The marks $VI\frac{1}{4}$ and $VI\frac{1}{2}$ in Mr. Lubbock's chart refer to changes in the current.

Mr. Dessiou's Tide Tables for Plymouth, &c., give the following tide times for places near Plymouth, deduced from Plymouth, of which the mean establishment is $5^h 33^m$. (p. 4.)

|  | h | m |
|---|---|---|
| Mount's Bay and Lizard . . . . . . | 4 | 30 |
| Falmouth Harbour . . . . . . . | 5 | 15 |
| Fowey Harbour . . . . . . . | 5 | 15 |
| Cawsand Bay . . . . . . . . | 5 | 23 |
| Eddystone . . . . . . . . . | 5 | 15 |
| Dartmouth and Torbay . . . . . . | 6 | 0 |
| Exmouth . . . . . . . . . | 6 | 25 |
| Lyme Cob . . . . . . . . . | 6 | 0 |
| Portland Bill . . . . . . . . | 5 | 30 |
| Weymouth . . . . . . . . . | 6 | 30 |

The last two are obtained from the same Tables, and deduced from Portsmouth, of which the establishment is $11^h 40^m$.

The Greenwich time at Portland Bill is hence $5^h 40^m$, and it is clear that the tides on each side of this point are bay tides. This is further confirmed by Mackenzie (Admiralty MSS.). He stationed observers at Portland Bill and at Weymouth, and found that high water at the Bill takes place above an hour sooner than at Weymouth.

As materials for the account of the tides on our own coast, we have many charts and surveys, executed by various persons, and especially by Murdoch Mackenzie and Græme Spence (1774 to 1792); and more recently a laborious survey by Captain Martin White, undertaken by direction of the Lords Commissioners of the Admiralty in 1812, of which the result has been printed, but is not yet published.

On the opposite side of the Channel we have,

| | h m | h m |
|---|---|---|
| Morlaix . . . . . . . . . . . | | 5 15 |
| St. Malo . . . . . . . . . . . | | 6 0 |
| Mont St. Michel . . . . . . . . | | 6 30 |
| Minquiers . . . . . | 6 0 | |
| Jersey . . . . . . | 6 10 | 6 0 |
| Guernsey . . . . . | 6 30 | 6 0 |
| Alderney . . . . . | 6 45 | |
| Cherbourg . . . . . . . . . . | | 7 45 |

The first column is from Captain M. WHITE's Survey of the Channel, p. 156; the second from the Annuaire du Bureau des Longitudes for 1833; the difference is not great. According to Captain WHITE's times, the cotidal line advances in a northerly direction out of the bay of Mont St. Michel, for the Minquiers rocks are to the south of Jersey, as Jersey is of Guernsey.

The longitude of Guernsey is $10^m$ W. Hence it would appear, that the $6^h$ $30^m$ cotidal line crosses the Channel about that part, and probably runs towards the east, so as to pass beyond St. Malo on the French coast.

This, however, is not what we should expect from another circumstance, the great height of the tides at the head of this bay, (at St. Malo and Granville they are forty or fifty feet,) which would lead us to suppose that there is a point of convergence in this neighbourhood.

The progress of the wave appears to be retarded by the opposite projections of Portland Bill and Cape La Hogue, especially the latter. The $7^h$ line appears to pass near the latter of these promontories.

Proceeding eastward, we have the following times, which I take from Mr. DESSIOU's Tide Tables for the English coast, and from the Annuaire for the French.

| | h m | | h m |
|---|---|---|---|
| Portland Bill . . . . | 5 30 | Le Havre . . . . . . | 9 15 |
| Needles Point . . . | 9 45 | Dieppe . . . . . . | 10 30 |
| Bembridge Point . . . | 11 0 | | |
| Shoreham Harbour . . | 11 15 | | |
| Beachy Head . . . . | 10 15 | | |

It appears that the $10^h$ $30^m$ line must pass nearly from Beachy Head to

Dieppe.   The inbend between Beachy Head and the Isle of Wight appears to have *bay* tides, which run up to Portsmouth.   The same is the case with respect to the inbend between the Isle of Wight and St. Alban's Head, which ends in Poole Harbour.

The following English tides are taken from Mr. DESSIOU's Sailing Directions for the English Channel, p. 53 and 60 ; the French, from the Annuaire.

|                     | h  | m  |                     | h  | m  |
|---------------------|----|----|---------------------|----|----|
| Rye Harbour . .     | 10 | 36 | Boulogne . . . . . .| 10 | 40 |
| Dungeness . . .     | 10 | 30 | Calais . . . . . . .| 11 | 45 |
| Folkstone . . .     | 10 | 45 | Dunkirk . . . . . . | 11 | 45 |
| Dover . . . .       | 10 | 50 | Ostend . . . . . . .| 12 | 20 |
| South Foreland .    | 11 | 0  |                     |    |    |
| Deal . . . .        | 11 | 15 |                     |    |    |
| Ramsgate . . .      | 11 | 20 |                     |    |    |
| North Foreland .    | 11 | 15 |                     |    |    |
| Margate Roads .     | 11 | 40 NORIE. |             |    |    |

The 11$^h$ 15$^m$ line passes from Deal to the nearest point of the opposite coast. The 12$^h$ line, which falls between Dunkirk and Ostend on the continental side, passes within the inlet of the Thames.

The tides beyond this point are affected by those which enter the German Ocean from the north, and must be considered hereafter.

### *Tides of St. George's Channel.*

I shall take these in the first place from Mr. DESSIOU's Sailing Directions. We have, on the opposite and parallel coasts of Cornwall and Devon and of Ireland, the following times, (pp. 124, 137.)

|                  | h | m  |                                  | h | m  |
|------------------|---|----|----------------------------------|---|----|
| Land's End . . . . . . | 4 | 30 | Cape Clear . . . . . . . .   | 4 | 0  |
| St. Ives Bay . . . . . | 4 | 30 | Cork and Kinsale Harbours .  | 4 | 30 |
| Padstow . . . . . .    | 5 | 0  | Between Waterford and Youghall | 5 | 0 |
| Lundy Isle . . . . . . | 5 | 15 | Hook Point (entrance of Water- | | |
| Barnstaple . . . . . . | 5 | 30 | ford) . . . . . . . .        | 5 | 15 |
| Ilfracombe . . . . . . | 5 | 30 | Saltee Islands . . . . .     | 5 | 50 |

2 A 2

Also we find that the line now reaches the promontory of the Welsh coast which divides the Bristol and St. George's Channels: for we have (p. 129),

|                   | h | m |
|-------------------|---|---|
| Milford Haven . . . . . . . . . . . . . | 5 | 30 |
| St. David's Head . . . . . . . . . . . | 6 | 0 |

I shall leave for the present the tides of the Bristol Channel, and follow the others.

The deep bay of Cardigan and Harlech, between St. David's Head and Bardsey Isle, will of course have *bay* tides. We have the following data (DES-SIOU, p. 137, 129.)

| | h | m | | | h | m |
|---|---|---|---|---|---|---|
| Wexford Harbour . . . | 7 | 0 | Fishguard Bay . . . | | 6 | 30 |
| Arklow. . . . . . . | 8 | 15 | Cardigan Bay . . . | | 7 | 15 |
| Wicklow . . . . . . | 9 | 0 | New Keyhead . . . | | 7 | 30 |
| | | | Aberystwith . . . . | | 7 | 45 |
| | | | Barmouth . . . . | | 7 | 45 NORIE. |

The lines are probably held back by the strait between the promontories of Carnsore and St. David's Head. In other respects they advance regularly to the north. The following Irish tides are from Captain MUDGE's Sailing Directions for Dublin Bay, p. 16; the English, from NORIE.

| | h | m | | | h | m |
|---|---|---|---|---|---|---|
| Kingstown Harbour . . | 10 | 17 | Bardsey Island . . . . | | 8 | 15 |
| Kish Bank Light . . . | 10 | 30 | Caernarvon Bar . . . . | | 9 | 0 |
| Balbriggen . . . . | 10 | 40 | Holyhead . . . . . | | 10 | 0 |
| Drogheda . . . . . | 10 | 40 | Amlwch Point . . . . | | 10 | 30 |
| Clogher Head . . . | 10 | 30 | Orme's Head . . . . | | 10 | 30 |
| Dundalk Bar . . . | 11 | 0 | Liverpool . . . . . | | 11 | 8 |
| Cooley Point . . . . | 10 | 40 | Lancaster . . . . . | | 11 | 15 |
| Carlingford Bar . . | 10 | 40 | Whitehaven . . . . | | 11 | 15 |
| Dundrum Harbour . . | 10 | 30 | St. Bees Head . . . . | | 11 | 0 |
| Strangford Bar . . . | 10 | 30 | Mull of Galloway . . . | | 11 | 15 |
| South Light . . . . | 10 | 15 | Mull of Cantire . . . . | | 10 | 30 |

It appears that there is a *point of convergence* somewhere near Dundalk, the cotidal lines north of this coming from the north. The general opinion, formed on the direction of the tide currents, that the "tides meet" at St. John's Point near Dundrum, agrees sufficiently well with this.

There is also a point of convergence on the English coast somewhere at the entrance of the Solway Frith: "the tides meet," according to the common opinion, at the Piel of Foudray, opposite the Isle of Man, where the tide has been known to rise thirty-six feet perpendicular, which is attributed to the accumulation of the two tides, (SPENCE, Scilly, p. 3.)

The Isle of Man shares the tides of the surrounding sea: we have in NORIE,

|  | h | m |
|---|---|---|
| Calf of Man | 10 | 30 |
| Douglas | 10 | 30 |
| Ramsey | 10 | 30 |

It appears, therefore, that the 11ʰ cotidal line passes beyond this island on both sides.

### Tides on the West of Ireland.

We must now trace the portion of the Atlantic tide, which, moving along the west coast of Ireland, comes round by the north and meets the tide of St. George's Channel. We have the following times at places along the west and north coasts, (NORIE).

|  | h | m |
|---|---|---|
| Skelling Rocks | 3 | 30 |
| Valentia Harbour | 3 | 30 |
| Dingle Bay | 3 | 45 |
| Tralee Bay | 3 | 45 |
| Loop Head | 4 | 30 |
| Galway Bay | 4 | 15 |
| Slyne | 5 | 15 |
| Achill Head | 6 | 0 |
| Donegal | 6 | 30 |
| Tory Island | 6 | 0 |
| Lough Swilly | 6 | 30 |
| Londonderry | 6 | 0 |

These times succeed each other very regularly, but there may be some doubt whether they are the results of exact observation at each place. In the Nautical Magazine for November 1832, is a survey by Captain MUDGE of the Gola

Islands, which are about twenty miles south of Tory Island; and there the time is given 4$^h$ 30$^m$. Also in Captain HUDDART's Survey of this coast, the time between Tory Island and Mallin Head (which is to the N.E.) is given as 5$^h$. I think it more probable, therefore, that 5$^h$ than that 6$^h$ is the time for Tory Island.

East coast.

|  | h | m |
|---|---|---|
| Fair Head | 9 | 0 |
| Carrickfergus | 10 | 30 |
| Belfast | 10 | 30 |
| Strangford Bar | 10 | 30 |

It appears that in passing round Fair Head, the north-east promontory, there is a *sudden* retardation of the wave, arising from the narrow channel between Fair Head and the Mull of Cantire, which is contracted still further by Rachlin Island.

This is further confirmed by the survey of that coast recently executed by Captain MUDGE; by which it appears that the tide time at the Giant's Causeway, a few miles to the west of this point, is 6$^h$; while at Torr Point, a little to the east of Fair Head, it is 9$^h$ 40$^m$. In going to the south, Captain MUDGE gives 10$^h$ as the time for Glenarm, 10$^h$ 5$^m$ for Belfast, and 10$^h$ 30$^m$ for Strangford, agreeing upon the whole pretty well with our previous data.

The suddenness of the change of the tide hour in passing round Fair Head is made more remarkable by the statement which Captain MUDGE gives of the tide in Ballycastle Bay, *immediately* to the west of Fair Head; he states for this place 5$^h$ 50$^m$, which is *earlier* than the tide at the Giant's Causeway to the westward.

This is in itself very extraordinary, and is opposed by the authority of the Survey published by Captain HUDDART (1790), who gives 7$^h$ as the tide-time in Ballycastle Bay. And I can hardly think Captain MUDGE's statement consistent with the time, 8$^h$, which he gives for Church Bay, on the opposite shore of the strait between Rachlin Island and the main land; a strait not above five miles across, including the bays. If both Captain MUDGE's statements are exact, the water must, for two hours ten minutes, be rising on one shore of this narrow channel, while it is falling on the other; an occurrence which seems scarcely possible.

It is to be observed, that the tide here is small (only four feet at springs) ; so that accuracy in observing the time of high water must be difficult to obtain. And as the currents between the island and the shore are very strong, it is possible that the change of level due to the shifting of currents may mask the regular tide so as to displace the time of high water by several hours.

Taking Captain Huddart's time for this point, it appears that for two hours forty minutes the water is falling in Ballycastle Bay within the Isle of Rachlin, while it is rising at Torr Point, which is beyond the Isle to the east. There must, therefore, be a strong current from the former to the latter of these points, which forms what is called the Race of Rachlin.

The retardation of the tide wave on the outside coast of Rachlin Island is probably not so abrupt as it is in the strait : still, however, it must be rapid ; and for reasons of the same kind as those just mentioned, we shall have a strong tide current between the island and the Mull of Cantire : this is called the Race of Skerinoe.

### Tides on the West Coast of Scotland.

The west coast of Scotland is so broken with promontories and islands, that the tides must be for the greater part *bay* tides; and it would introduce confusion to attempt to trace them in a general survey like the present. I shall therefore only consider the course of the main wave.

By Captain Huddart's map it appears, that the 5$^h$ line (or for *Greenwich time* the cotidal line of 5$^h$ 30$^m$,) passes near Tory Island. The same line, it appears, bends towards the shore so as to approach Icolmkill or Mull, for which also 5$^h$ is marked. To the S.E. of this line, as at Isla, &c., the tides are later. The southern extremity of the range called the Western Isles is marked 5$^h$ 30$^m$, which is also the time given for the strait between North Uish and Harris. (Mr. Lubbock gives 6$^h$ 30$^m$.) Inside of the Lewis we have 6$^h$, both on the side of the island and the main land; and further north in the Minsh, near Cape Wrath, we have 7$^h$. Norie gives 8$^h$ 15$^m$, which, like his other statements for this western coast, is considerably larger than our authorities give.

For the Orkney Isles I find the following statements in Norie's Sailing Directions for the East Coast of England and Scotland, p. 55.

|                                               | h | m |
|-----------------------------------------------|---|---|
| South Ronaldsha                               | 9 | 0 |
| Stromness, Wells, and Westra                  | 9 | 0 |
| Foul Island                                   | 9 | 30 |
| Fair Island                                   | 10 | 0 |
| Brassa Sound                                  | 9 | 45 |
| East side of Sanda and North Ronaldsha        | 9 | 45 |
| North Ronaldsha Frith                         | 10 | 45 |
| East side of Shetlands                        | 9 | 30 |
| Duncansby Head                                | 8 | 15 |

It is tolerably evident from this, that the $9^h$ tide line nearly (Orkney time,) is the one belonging to these islands, where the wave is not retarded by narrow channels.

### Tides of the Arctic Ocean.

The main tide-wave which we have been following, after reaching the Orkneys, will, I conceive, move forwards in the sea of which the shores of Norway and Siberia form one side, and those of Greenland and America the other. It will here meet in succession with the islands of Iceland and Spitzbergen*; it will pass the pole of the earth, and will finally end its course on the shores in the neighbourhood of Behring's Straits. Perhaps it may propagate its influence through the Straits, and modify the tides of the North Pacific.

But a branch tide is sent off from this main tide into the German Ocean: this, entering between the Orkneys and the coast of Norway, brings the tide to the east coast of England, the coast of Holland, Denmark, and Germany. We shall now trace this tide.

---

* I add the following statements, which refer to the further course of the tide.

| | h | m | |
|---|---|---|---|
| Bergen, lat. 60° 24' | 1 | 30 | Norie. |
| Drontheim, lat. 63° 26' | 2 | 15 | —— |
| Hammerfest, lat. 70° 40' | 1 | 10 | —— |
| North Cape, lat. 71° 10' | 3 | 44 | —— |
| Sweetnose (Lapland), lat. 68° 10' | 8 | 30 | —— |
| Isle Kilduin (Lapland), lat. 69° 10' | 7 | 30 | Lalande, p. 340. |
| Archangel | 6 | 0 | ——, pp. 272 & 340. |
| Patrix Fiord (Iceland), lat. 65° 36' | 6 | 0 | ——, p. 340. |
| Hakluyt's Head (Spitzbergen) | 1 | 30 | beginning of tide. Phipps, pp. 44 & 67. |
| Magdalen Bay | 1 | 30 | Phipps, p. 30. |
| Moffen Isle, 25th July 1773, low water at | 11 | 0 | ——, p. 50. |

*Tides on the East Coast of Britain.*

Taking for our authority NORIE's Sailing Directions for the East Coast, we find the following tide-hours (p. 38):

|  | h | m |
|---|---|---|
| Duncansby Head | 8 | 15 |
| Sinclair's Bay | 9 | 0 |
| Frith of Tain | 11 | 0 |
| From Bamff to Cromarty | 11 | 45 |
| *Inverness* | 12 | 0 |
| Buchanness | 12 | 0 |

It appears that Inverness has a bay tide, a point of divergence occurring on the shore of Bamff.

By comparing the time given for Duncansby Head with that given for the Orkneys, and with that for Sinclair's Bay, which is a very little to the south of it, we should be led to think it earlier than the true time.

We have also (p. 32) the following tide-hours:

|  | h | m |
|---|---|---|
| Newburgh | 12 | 30 |
| Aberdeen | 12 | 45 |
| Montrose and Stonehaven | 1 | 30 |
| *Tay Bar* | 1 | 45 |
| *Dundee* | 2 | 15 |
| *St. Andrew's* | 2 | 0 |
| Fifeness | 1 | 30 |
| *Leith* | 2 | 20 |
| Dunbar | 1 | 30 |

The places in Italics are affected by the retardation produced by the Friths of Tay and Forth.   Perhaps Montrose is partly affected by this cause.

We proceed southwards, and find (pp. 17 and 2),

|  | h | m |
|---|---|---|
| Berwick and Eymouth | 2 | 15 |
| Holy Island | 2 | 30 |
| Fern Island | 2 | 40 |
| Blyth and Coquet Island | 2 | 45 |
| Tynemouth Bar | 2 | 50 |

|                                                    |  h |  m |                  |
|----------------------------------------------------|---:|---:|------------------|
| Sunderland . . . . . . . . . . . . .               |  3 |  0 |                  |
| Hartlepool, mouth of the Tees, and Whitby          |  3 | 30 |                  |
| Scarborough . . . . . . . . . . . .                |  4 | 15 |                  |
| Flamborough Head . . . . . . . .                   |  4 | 30 |                  |
| Bridlington Bay . . . . . . . . . .                |  4 | 30 |                  |
| Spurn Point . . . . . . . . . . . .                |  5 | 15 |                  |
| Sandhale . . . . . . . . . . . . .                 |  6 |  0 |                  |
| *Hull Road* . . . . . . . . . . . .                |  6 | 15 |                  |
| Dudgeon Shoal (at the mouth of the Wash)           |  6 |  0 |                  |
| *Lynn Well* . . . . . . . . . . . .                |  6 | 30 |                  |
| Cromer . . . . . . . . . . . . . .                 |  7 |  0 | NORIE's Epitome. |
| Lowestoff Roads . . . . . . . . .                  |  8 | 55 | ———————          |
| ————————, in shore . . . . . . .                   | 10 | 38 | ———————          |
| Harwich . . . . . . . . . . . . .                  | 11 | 30 | ———————          |

As this time agrees with that at the North Foreland nearly, we may suppose the tide-wave to extend from one point to the other, and the tide which runs up the Thames from this point is a river-tide.

### Tides on the remaining Coasts of the German Ocean.

We have the time at the Naze of Norway given as $11^h 15^m$, which corresponds with that on the shore of Caithness, so that the tide-wave would at first appear to extend directly across the northern opening of this sea. We find, however, that this cannot be the course of the cotidal lines, for we have $11^h$ for the hour at Heligoland, and hours a little later for the adjacent coasts; namely, (Sailing Directions for Heligoland, p. 8; and NORIE,)

|                                            |  h |  m |
|--------------------------------------------|---:|---:|
| Scaw (north point of Jutland) . . . . . . . . . . . . | 12 |  0 |
| Hoorn (west coast of Denmark) . . . . . . . . . | 12 |  0 |
| Island of Heligoland . . . . . . . . . . . . | 11 |  0 |
| Red Buoy (entrance of the Elbe) . . . . . . . | 12 |  0 |
| Cuxhaven . . . . . . . . . . . . . . . . . . . | 1 |  0 |
| Island of Wrangeroog (mouth of the Weser) . . . | 12 |  0 |
| Island of Borkhum (mouth of the Ems) . . . . | 11 | 30 |
| Emden . . . . . . . . . . . . . . . . . . | 12 |  0 |

Also, proceeding to the coast of Holland, we have (Directions for the North Sea,) these tide-hours:

|  | h | m |
|---|---|---|
| Island of Ameland . . . . . . . . . . . . . | 10 | 30 |
| Vlie Passage . . . . . . . . . . . . . . . . | 9 | 0 |
| Texel (entrance to) . . . . . . . . . . . | 6 | 45 |
| Texel Road . . . . . . . . . . . . . . . | 7 | 45 |
| Camperdown . . . . . . . . . . . . . . | 4 | 30 |
| Hook of Holland . . . . . . . . . . . . . | 3 | 0 |
| Briel (mouth of the Maes) . . . . . . . . . | 3 | 0 |
| Browershaven (East Schelde) . . . . . . . . . | 2 | 0 |
| Veer . . . . . . . . . . . . . . . . . . | 1 | 20 |
| Flushing (West Schelde) . . . . . . . . . . | 1 | 20 |
| Blankenberg . . . . . . . . . . . . . . | 12 | 20 |
| Ostend (as before) . . . . . . . . . . . . | 12 | 20 |

It appears quite clear from this list, that the tide-wave on this coast runs to the east; and this is also the case with the tide currents; the general flood-stream running N.E., and the ebb S.W. (Sailing Directions, p. 55.)

We seem to be led therefore to the remarkable conclusion, that a tide-wave which reaches almost or quite to the shores of Denmark, proceeds through the narrow opening of the Straits of Dover, though the main part of the German Ocean is occupied with the tide-wave which enters by the north; and that the tide-wave on the opposite shores of England and Holland runs at the same moment in opposite directions.

In consequence of this state of things, it is very difficult to assign the form and motion of the tide-wave in the middle parts of this ocean. The tides there will be *tides of interference;* (see Sect. 1.), and the undulations of the surface may be stationary instead of successive undulations, and may be defined by spaces instead of lines. Numerous observations at a distance from shore will be requisite to determine the circumstances of these undulations.

On the Leman and Ower shoals, $6^h 30^m$ or $7^h$ is the tide-hour. The flood comes from the north upon the whole, though its direction varies through a certain angle during the flow; whether the flood sets north or south off the coast of Denmark, I do not find stated. Captain ANDERSON (Phil. Trans.

1819, p. 226,) says, "The opposite tides which meet in the North Sea do not meet in a line directly across any part of it, but in a diagonal line extending from the Kentish Knock to the entrance of the Sleeve;" which seems to imply that the tide on the Danish coast comes from the S.W. He adds, "In fact, there is hardly any tide observable between the Hoorn reef and the entrance of the Sleeve."

The $7^h$ tide line on the Ower shoal is further back than the same line on the shore; which, being contrary to the course of single tides, appears to indicate there a tide of interference; the absence of tide on the coast of Jutland is also probably the result of interference, for the tides at Heligoland are said to rise 9 feet (NORIE). If we suppose the southern and northern tides to reach this space at an interval of 6 hours,—for instance, if the tide-hours for these tides be respectively $3^h$ and $9^h$,—the tide will be altogether obliterated by their combination.

We may therefore account for several of the facts, by supposing the $9^h$ tide-line to advance from the north in a convex form, so as to approach this part; and the southern tide to move so as to produce the lines $12^h$, $1^h$, $2^h$, $3^h$ on the coast of Denmark, if it were single. We may thus suppose in the middle of the ocean *a space* which has its tide at $6^h$ nearly, by the interference of the tides

at     $4^h$ $5^h$ $6^h$ $7^h$ $8^h$ from the north,

and    $3^h$ $7^h$ $6^h$ $5^h$ $4^h$ from the south.

A circumstance in the rise of the tide in this part appears to confirm the opinion that it is a tide of interference. "A singular peculiarity in the tide about the Ower was observed;—there was no sensible *rise* in the tide until 3 hours after low water; and when the ebb stream was nearly done, a sudden rise of 5 or 6 feet took place; so that nearly the whole rise of the tide occurs in the last 3 hours of it." (Captain HEWETT on the Leman and Ower Shoals.)

Still it is scarcely possible that a tide so considerable as that on the north coast of Germany should be entirely produced by the undulation propagated through the Straits of Dover. By looking at the map, it will be seen that if we draw a line from Norfolk on one side to the Texel on the other, we cut off a portion of the sea extending to Dover, which may be considered as merely an inlet of the German Ocean; and the tides of the main expanse might be expected to be nearly independent of those of this inlet. If this inlet were closed, the tide-wave would pass from the coast of Norfolk to that of Holland, the hour being 6$^h$ in Lynn Wash, and 7$^h$ at the Texel; proceeding regularly from Caithness to Heligoland.  This is a view of the origin of the tides of the German Ocean, different from the former one.  At present it appears to be difficult to decide which is the more correct.

### On the Tides of the South Atlantic.

It will be recollected that the tide which entered the South Atlantic found a point of divergence at or near Cape Frio on the coast of Brazil.  We have traced the northern branch of the tide to the end of its career, and have particularly examined its movement on our own and the neighbouring coasts. We shall now endeavour to trace the southern branch of this tide.

Proceeding along the coast southwards from Cape Frio, we find Rio Janeiro at a small distance, for which we have the establishment 2$^h$ 40$^m$ (NORIE), or 2$^h$ 45$^m$ ROUSSIN; and adding 2$^h$ 52$^m$ for longitude to the latter, the tide-hour, Greenwich time, is 5$^h$ 37$^m$: Mr. LUBBOCK gives 5$^h$ 15$^m$.

The island of Santa Catharina, in south latitude 27°, is the next point. NORIE gives 2$^h$ 30$^m$; ROUSSIN 2$^h$ 45$^m$, or Greenwich time 5$^h$ 57$^m$, which agrees nearly with Mr. LUBBOCK.

In the Sailing Directions for the coast of Brazil which I have seen, I do not find any account of the tide-hours at any point of the coast south of Santa Catharina*.  Our next materials are Captain KING's survey of Patagonia.

In Captain KING's Sailing Directions for the Coast of Eastern and Western Patagonia, we have for Port St. Elena, in latitude 44° 30′, the time of high water at full and change 4 o'clock.  The longitude is 4$^h$ 20$^m$ W.  Hence the tide-hour, Greenwich time, is 8$^h$ 20$^m$.

It would at first appear from this, that the tide-wave employs less than 4 hours in passing from Cape Frio to Port St. Elena, running southwards along the shore: but we shall find various circumstances which will not allow us to consider this as the true motion of the tide-wave.  It appears from Captain KING's observations (Directions, p. 17), that along the whole of the coast in

---

* The tide at Monte Video is said to be at noon (Forte's Remark Book), and at Blanco Bay at 6$^h$; it is easy to accommodate the cotidal lines to the latter datum; the former is probably affected by the influence of the river.

the neighbourhood of Port St. Elena, the flood tide sets to the *northward*, while off the shores of Brazil the flood sets to the *southward*, (NORIE's Sailing Directions for Brazil, Part I. p. 45). And though it is not universally necessary that the cotidal lines should move in the same direction as the flood stream, there can hardly be a discrepancy between these directions on such a scale as the southward motion of the tide-wave to St. Elena would seem to imply. We are therefore compelled to suppose that somewhere on the coast to the north of St. Elena there is *a point of convergence of cotidal lines,* at which point they are later than either to the north or to the south of it.

This view is confirmed by our finding that on the coast of Patagonia to the south of Port St. Elena, the cotidal line appears undoubtedly to travel northwards. Thus Captain KING, p. 17, makes the following statement, taking points from Cape Virgins at the eastern entrance of the Strait of Magelhaens, in order, going northwards.

| | Latitude. | Tide-hour. | Feet. |
|---|---|---|---|
| | | h  m | |
| Cape Virgins . . | 52 21 S. . . | .about 8  0 | |
| Gallegos River . . | 51 43 . . | . . . 8 30 . . . | . rise 46 |
| Cape Fair Weather | 51 33 . . | . . . 9  0 . . . | . . 28 |
| Coy Inlet . . . . . . | | . . between 9 and 10h | |
| Santa Cruz . . . | 50 17 . . | . . . 10 15 . . | . . . 33 |
| Port St. Julian . . | 49  8 . . | . . . 10 34 . . | . . . 38 |
| Sea Bear Bay . . | 47 56 . . | . . . 12 45 . . | . . . 20 |
| Port Desire . . . | 47 45 . . | . . 1  0 . . | . . .21½ |
| Port St. Elena . . | 44 30 . . | . . . . 4 "in the afternoon." | |

Hence it appears that the tide travels from Cape Virgins northward to Port St. Elena in about eight hours: and as, from the form of the coast, we cannot doubt that the tide must reach Cape Frio before it penetrates the bay formed by the east coast of Patagonia, it appears that the tide at Port St. Elena is not four hours, but sixteen hours, after that at Cape Frio.

The tide which comes from the southward is one of great magnitude, the spring tide heights ranging from twenty to thirty feet along the coast, and attaining forty-six feet at the Gallegos River. The tide which comes from the north is much smaller; it is stated at five feet at Cape Frio, four feet at Rio, six feet at Santa Catharina. In the River Plata the tides do not rise or fall more than five or six feet (Captain HEYWOOD in NORIE's Brazil, p. 47). Hence the tides to the south of the River Plata must be principally produced by the wave which comes from the south. Probably Cape Corrientes, in latitude 38°, may be taken as the southernmost headland reached by the tide from the north. We have no evidence (so far as I am aware,) what time the northern tide employs in reaching this part of the shore. I shall suppose Cape Corrientes to be the point at which the tide is simultaneous with that at Cape Virgins; and between these points the cotidal lines will advance from both extremities to an intermediate *point of convergence of cotidal lines,* which may perhaps be at the heads of the Gulf of St. George, or of St. Antonio, in latitude 41° and 46°.

Though this determination of the form and course of the cotidal lines of this coast may be considered somewhat hazardous, it derives considerable confirmation from what takes place on another coast somewhat similarly circumstanced, that of North America; besides being, so far as I can perceive, the only way in which the observations can be made consistent with each other.

The cotidal lines which are drawn in the chart may appear very much *thronged;* but it may be observed that we have good evidence that they must be as close to each other as they are there drawn, because there are eight of them which reach the shore between Cape Virgins and Port St. Elena, about 8° of latitude. We have also from PURDY (E. M. p. 59,) the following statement of the times of high water at various points of the Falkland Islands:

|  | h | m |
|---|---|---|
| Berkeley Sound | 5 | 0 |
| North entrance of Falkland Sound | 6 | 30 |
| Tamer Harbour | 7 | 0 |
| *Pebble Sound* | 8 | 30 |
| Saunders Island | 7 | 30 |
| Jason's Isles | 8 | 0 |
| Swan Islands | 8 | 30 |
| Port Stephens | 8 | 0 |
| Port Albemarle | 7 | 45 |

Swan Islands and Port Stephens are on the western side of this land, and therefore retarded; and as 4ʰ W. is the longitude of this land, the cotidal line of $9^h$ appears to pass near the eastern part of these islands; while at Cape Virgins, in longitude $4^h 33^m$ W., we have the line of $12^h 30^m$.

The cotidal line at the eastern extremity of Tierra del Fuego is that of $8^h 35^m$ Greenwich time; for in Good Success Bay, latitude 54° 48′ S., longitude $4^h 20^m$ W., the time as stated by Captain KING is $4^h 15^m$.

The time of high water at South Georgia (COOK in PURDY's E. M. p. 36,) is $11^h$; and as the longitude of this land is about $2^h 24^m$, the cotidal line of $1^h 30^m$ passes near it.

### On the Tides of the Pacific.

1. *Western coast of America.*

From Cape Pillar, at the western extremity of the Strait of Magalhaens, the shore of Tierra del Fuego runs E.S.E. to Cape Horn, and then E.N.E. to Strait le Maire and Staaten Island. Along this coast the tidal wave travels to the *eastward*. Thus, according to Captain KING (Sailing Directions, p. 96, and Table, pp. 13, 14,) at Cape Pillar it is high water at $1^h$ on the days of full and change; at York Minster, 5° of longitude to the east, it is at $3^h$; at Cape Horn, 3° further east, it is at $3\frac{1}{2}^h$; in Good Success Bay, in Strait le Maire, the hour is 4; on the east side of Strait le Maire it is $5^h$, (p. 100). As there is a constant easterly current off Cape Horn (p. 102), the ebb may be supposed to be produced by a retardation only of this motion: nearer Cape Pillar, the tide current came from the N.W. (p. 96); the tides here are about four feet.

The motion of the cotidal lines upon a large scale is undoubtedly from east to west: and it is certainly very remarkable that in this part of the ocean where there appears to be nothing to interrupt the westerly motion, the lines should travel in an opposite direction. This however is only a partial phenomenon, like the convergencies on various coasts which we have already had to notice; for a little further to the east and north, at Staaten Island, and on the coast north of Cape Diego (at Strait le Maire), the tides set to the north and west (p. 106).

In order to determine the motion of the tide-wave on the west coast of America, we have the following data:

| | Latitude. | Tide-hour. | Green. Time. | |
|---|---|---|---|---|
| | °　′ | h　m | h　m | |
| Cape Pillar .............. | 52　46 S. | 1　0 | 6　0 | |
| San Carlos de Chiloe ...... | 41　52 | 11　30 | 4　26 | 6ft. King's Table, p. 15. |
| | | 12　30 | ...... | 12ft. Heron R. B. |
| Valdivia ............... | 39　50 | 11　30 | { 4　24 | Norie; Purdy, E. M. p. 59. |
| | | | { 4　45 | Lubbock. |
| Conception ............. | 36　49 | 10　0 | 2　50 | Malaspina, p. 127. |
| | | 8　30 | ...... | Thetis R. B. |
| | | 9　45 | ...... | Thetis R. B. |
| *Conception* .............. | ...... | [1　30] | ...... | Forte R. B., and Perouse in Purdy E. M. p. 204. |
| Talcahuana (harbour of Conception) .............. | ...... | 10　0 | ...... | Bauza, MS. |
| *Talcahuana* .............. | ...... | [3　20] | ...... | Beechey, p. 645. |
| Valparaiso ............. | 33　2 | 9　25 | 2　10 | Malaspina, p. 127. |
| | | 9　0 | | Tribune R. B. |
| Coquimbo ............. | 29　54 | 9　0 | 1　45 | Tribune R. B. |
| | | 9　40 | | Forte R. B. |
| *Coquimbo* ............... | ...... | [2　37] | ...... | Malaspina, p. 127. |
| *Copiapo* ............... | 27　12 | [2　30] | ...... | Lubbock. |
| Callao ................. | 12　4 | 6　15 | 11　23 | Malaspina, p. 127. |
| Guayaquil ............. | 2　12 | 7　19 | 12　42 | Ibid. |
| Galapagos ............. | 1　21 | | | |
| ——— Charles's Bay .... | 1　0 | 2　0 | 8　1 | |
| ——— Chatham Isle .... | 1　0 | 3　30 | ...... | Purdy, E. M. p. 51. |
| Cocos ................. | 5　34 N. | 2　10 | ...... | ———, E. M. p. 50, Vancouver. |
| | | 4　0 | ...... | ———, E. M. p. 47, Colnett. |
| Panama ............... | 8　57 | 2　30 | 7　47 | Lloyd in Phil. Trans. 1830. |
| | | 2　0 | ...... | Foster (S. A. P. 2, p. 91.) |
| Realejo ............... | 12　30 | 2　43 | 8　31 | Malaspina, p. 127. |
| Acapulco ............... | 10　50 | 1　19 | 7　59 | Ibid. tide very small. |

Among the different hours given for Conception, there can be little doubt that 1$^h$ 30$^m$ is erroneous, whether the error arise from the confusion of high water with slack water, or from whatever other cause; the superior weight of evidence and the times of neighbouring places show that the time here must be about 9$^h$, though Lieutenant Beechey's statement is so different. Similar considerations induce us to reject 2$^h$ 37$^m$ for Coquimbo*. And the causes, whatever they are, which have given rise to these erroneous statements for Conception and Coquimbo, have probably given rise to the statement for Copiapo, which Mr. Lubbock has recorded, and which is irreconcilable with the course of the tides on the rest of the coast.

* On referring to the original MS. of the observations made in Malaspina's voyage, which Mr. Bauza has obligingly shown me, I find that while the time of tide at the other places is termed " plea-mar," at Coquimbo it is stated as " hora en que cambio."

Omitting, then, the times thus condemned, we perceive that there are, with slight irregularities only, hours perpetually earlier and earlier as we go further north all the way to Acapulco, and that therefore the tide-wave travels along this coast from north to south, employing about twelve hours in its motion from Acapulco to the Straits of Magalhaens.   In confirmation of this view it may be observed, that according to Captain COLNETT (PURDY, E. M. p. 216), at the island of Quibo, westward of the Bay of Panama, the flood comes from the north at full and change, " flowing seven hours and ebbing five :" and along the whole coast of Peru and Chili, the tide sets to the southward (LALANDE, p. 291).

At the Gallapagos, according to PURDY (E. M. p. 59), the flood is from the east; as it also is at the Cocos (p. 50).

The Gulf of Panama has nearly the same tide-time as the Gallapagos and other islands opposite to it in the open ocean.   When we consider the deep inbend of the shore of this bay, and the great height of the tides which take place in it (eighteen or twenty feet at St. Michael's Bay on the eastern side of the Gulf, LALANDE, p. 293), it is difficult to suppose that the tide-wave occupies so short a time in travelling up the Gulf as this statement would imply.   If, therefore, the hours given for Cocos Island, Charles's Island, and Chatham Island are nearly right, it is more probable that the tide-wave employs twelve hours in passing from the neighbourhood of the Gallapagos to the extremity of the Gulf.   The tide-line will travel slowly in the narrower part of the bay, especially if it be shallow ; and I shall adopt this as the most probable view of the cotidal lines, till we have more numerous and more certain observations.

As we proceed further north we have evidence that the tide-line is now travelling north : we find

| | Latitude. | Tide-hour. | |
|---|---|---|---|
| | ° ′ | h  m | |
| San Blas .................. | 21 32 | 8  5 | Mem. on S. Am. P. 2. p. 100. |
| Mazatlan.................. | 23  0 | 9 41 | BEECHEY, p. 661. |
| | | 9 50 | ————, p. 661. |
| San Diego ............... | 32 42 | 9  0 | |
| San Carlos de Monterey  .... | 36 36 | 10  0 | Mem. on S. Am. |
| | | 9 42 | BEECHEY, p. 655. |
| | | 8  0 | MALASPINA. |
| San Francisco.............. | 37 48 | 10 52 | ———— |
| *Colombia* R................. | 46 19 | 1 30 | |
| Nootka Sound.............. | 49 34 | 0 20 | NORIE. |

The tide at the Colombia is probably retarded by the inlet.

It would seem therefore that there must be a point of divergence somewhere in the neighbourhood of Acapulco (lat. 16° 50′). The tide at Acapulco is difficult to reconcile with the rest; and as the rise is very small, we may perhaps venture to suppose the time considerably erroneous.

2. *Central parts of the Pacific.*

I fear it is at present impossible to draw the cotidal lines, with any strong probability of their being right, between the coasts of Chili and New Zealand. We should certainly expect them to travel from east to west; and as there must, as we should conceive, be twenty-four different cotidal lines in going round the globe, there must, it would seem, be a considerable number of them in this wide ocean. Our knowledge of the tides in this part of the world is very scanty. The extreme smallness of the height of the tide makes it difficult to observe it with certainty, and allows it in many cases (for instance, in the Sandwich Isles,) to be completely masked by the constant daily effect of the land and sea breezes.

Proceeding westward across the Pacific, we have the following statements:

| | Long. W. | Tide-hour. | Green. Time. | |
|---|---|---|---|---|
| | h   m | h   m | h   m | |
| Gallapagos Island, Charles's Bay. | 6   1 | 2   0 | 8   1 | 7 to 8 feet. |
| Easter Island ................ | 7  18 | 2   0 | 9  18 | Norie. |
| Gambier's Group ........... | 9   0 | 1  50 | 10  50 | Beechey, p. 646. |
| Lagoon Island................ | 9  18 | 0  30 | 9  48 | Cook, Phil. Trans. 1772. |
| | | 11  15 | 8  33 | Lalande. |
| Society Islands. | | | | |
| Otaheite .................... | 9  58 | 0  15 | 10  13 | ———— |
| Ulietea...................... | 10   6 | 11  30 | 9  36 | ———— |
| Huaheine (Owharree Bay)...... | 10   4 | 11  50 | 9  54 | Norie. |

The times for Otaheite and Ulietea are calculated (Lalande, p. 298,) from observations made for several days. The tides in this group of islands are extremely small; at Otaheite, for instance, only eleven inches. This circumstance is attributed, by Mr. Wales, to the banks of coral which surround these islands, and which leave only narrow openings for the tide to pass in and out: the same cause will probably affect the time of high water, but it is difficult to say how much. Indeed, we may doubt whether this small rise really exhibits the effect of the lunisolar tide. In some places other diurnal influences completely mask the effect of the attraction of the sun and moon. Thus, Captain Beechey observes, that at Papiate, one of the Society Islands, it is high water *every day* at half an hour past noon, and low water at six in the evening.

Lieutenant MALDEN (Lord BYRON's Voyage,) gives a similar account of the tides at Owhyhee.

The establishment for Resolution Bay in the Marquesas Islands (long. W. $9^h 15^m$) is stated at $2^h 30^m$, which seems to show some irregular form of the tide-wave. The deficiency of materials, however, with regard to this part of the sea, is so great, that I shall not attempt to draw the cotidal lines in the Pacific north of the equator.

3. *Western parts of the Pacific.*

When we arrive at the more extensive and closer islands on the west side of the Pacific, we can trace the course of the tides with a little more definiteness. Captain COOK has stated the times for several points on the coast of New Zealand (Phil. Trans. 1772.). At Tolaga Bay, near the most easterly point of these islands, the time is $6^h$. In proceeding to Mercury Bay and the Bay of Islands, on the N.E. coast, it becomes $7^h 30^m$ and $8^h$ respectively. Again, in proceeding southward from Tolaga Bay, we have also a retardation. At Queen Charlotte's Sound and Admiralty Sound, in Cook's Strait which separates the two islands, it is $9^h 30^m$ and $10^h$, the Strait producing a considerable retardation. At Dusky Bay, at the southern point of the land, the time is $10^h 57^m$. Hence it is clear that the $6^h$ cotidal line meets the coast near Tolaga Bay, and there forms a point of divergence, the tides passing round the ends of the islands to the north and south, occupying three or four hours in turning each extremity.

The tide which was on the coast of Patagonia at five o'clock, and on the coast of New Zealand at six, must have employed thirteen hours in this passage; but, as it is moving to the southward and eastward in the former case, and to the westward in the latter, the cotidal lines must somewhere have a vertex, which is probably turned towards the north.

The Friendly Islands, which are nearly to the north of New Zealand, receive the tide-wave about the same time as that coast, as appears by the following data:

|  | Longitude. h   m | Tide-hour. h   m | Green. Time. h   m |
|---|---|---|---|
| New Zealand. |  |  |  |
| Tolaga Bay ............ | 11  53 E. | 6   0 | 6   7 |
| Friendly Islands. |  |  |  |
| Annamooka ............ | 11  40 W | 6   0 | 6  20 |
| Tongataboo ............ | 11  41 | 6  50 | 7  19 |
| Eooa................ | 11  40 | 7   0 | 7  20 |
| Wallis's Island.......... | 11  44 | 5   0 | 4  44 Zebra R. B. |

Hence it appears probable that the 6$^h$ and 7$^h$ cotidal lines extend nearly north and south as far as the equator; and the following data, referring to places which lie in the intermediate space, but a little more to the west, agree with this view.

|  | Longitude.<br>h  m | Tide-hour.<br>h  m | Green. Time.<br>h · m |  |
|---|---|---|---|---|
| Norfolk Island.......... | 11  12 E. | 7  45 | 8  33 | NORIE, LUBBOCK. |
| New Caledonia. |  |  |  |  |
| Balada Harbour ........ | 10  58 | 6  30 | 7  32 | ———— |
| Pudyona .............. | .......... | 6  30 |  | ———— |
| New Hebrides. |  |  |  |  |
| Tanna, Port Resolution .. | 11  19 | 5  45<br>[3   0] | 6  26 | LALANDE, p. 298.<br>NORIE. |

The tide occupies an hour or two in passing to the other coast of New Caledonia around the north or south promontories; at Port St. Vincent, on the west shore, the hour is 8$^h$ 10$^m$ (NORIE).

Three or four degrees south of the south point of New Zealand are Lord Auckland's Isles, discovered by Capt. BRISTOW in 1806. Mr. LUBBOCK states the tide hour at 11$^h$ 30$^m$, which, if true, indicates that the cotidal lines here are brought near each other even at a considerable distance from the coast of New Zealand. It is difficult to conceive that the tide at a detached island should be half an hour later than at Dusky Bay, which is not only more to the west, but has the tide brought to it by a wave which travels along an extensive retarding coast, round a long promontory, and through a large portion of a whole circumference.

### On the Tides of the Coasts of Australia.

For the tides of the Australian coasts our materials are more abundant, Capt. FLINDERS and Capt. KING having pretty fully examined these seas. The most easterly part of the shore, from lat. 24° to 35° S., appears to receive the tide earliest. Thus we have,

|  | Lat.<br>°  ′ | Tide-hour.<br>h  m |  |
|---|---|---|---|
| Bustard Bay  .  .  . | 24  30 | 8   0 | COOK, Phil. Trans. 1772. |
| Hervey's Bay  .  .  . | 24  40 | 8   0 | FLINDERS (ii. 11). |
| Botany Bay  .  .  . | 34   0 | 8   0 | COOK, FLINDERS. |

Various points within this district have the tide later, but they are places where a promontory or island intervenes between them and the open sea. Thus,

| | Lat. | Tide-hour. |
|---|---|---|
| | | h   m | |
| Port Jackson . . . | 33 50 S. . . . | 8 15 FLINDERS. |
| Rodd's Bay . . . | 23 59 . . . | 8 30 KING (ii. 261). |
| Shoal's Haven . . . | 34 45 . . . | 8 30 FLINDERS. |
| Moreton Bay . . . | 27 10 . . . | 9 30 ——— |

In proceeding to the northwards, we find that the hour becomes perceptibly later. We have,

| | Lat. | Tide-hour. |
|---|---|---|
| | | h   m |
| Port Curtis . . . . . . . | 23 52 S. | . 8 to 9   0 FLINDERS. |
| Keppel Bay . . . . . . | 23  8 . . . | 9 30 ——— |
| Port Bowen . . . . . . | 22 28 . . . | 10  0 ——— |
| Strong-tide Passage . . . . . . . | | 10  0 ——— |
| Shoal-water Bay . . . . . . . . | | 10 30 ——— |
| Thirsty Sound . . . . | 22  6 . . . | 10 45 ——— |
| Broad Sound . . . . . . . . . | | 11  0 ——— |
| Percy Islands . . . . | 21 19 . . . | 8  0 ——— |
| Cape Hilsborough . . . | 20 53 | |
| Cumberland Isles . . . . . . . . | | 11  0 ——— |
| Endeavour River . . . . | 15 27 . . . | 9 30 ———, COOK. |
| Princess Charlotte's Bay . . . . . . . | | 8  0 KING (ii. 281). |
| Endeavour's Strait . . . . | 10 37 . . . | 1 30 ——— |
| Murray's Island in Torres' Strait . . . . . } | . 9 55 . . . | 10 30 FLINDERS. |

The earliness of the tide at Murray's Island may be accounted for, as the island is at some distance from the main shore, and the tide-line will, as usual, be convex. The tides at Percy Islands are less consistent with the rest. Capt. FLINDERS (ii. 82.) gives the time as observed by Capt. FOWLER, and expresses his surprise at the difference of three hours from Broad Sound. Capt. KING (ii. 263.) observed the tide on two successive days to be about 10 hours, and 11 hours after the moon's transit. This, even corrected for the semi-menstrual inequality (see Sect. 2), would give the hour much nearer that on the shore.

The earliness of the time given by Capt. KING for Endeavour River, compared with the times at adjacent places, and its differing both from FLINDERS and COOK, are grounds which authorize us to suspect it.

Hence we collect that the tide-wave occupies about 3 hours in moving northwards along the coast from lat. 30° to Torres' Straits between New Holland and New Guinea, and reaches the points within the Straits westward later still. The cotidal line which meets the coast at $8^h$, in long. $10^h 15^m$ E., is that of $9^h 45^m$; and the one at the Straits, in long. $9^h 30^m$ E., is that of $1^h$. This agrees with Capt. KING's statement (ii. 259), that in this part the flood sets N.W.

Beyond this point the general run of the coast is to the west; and we can trace the course of the tide by means of Capt. KING's Survey. We have the following data.

| | Long. E. h m | Tide-hour. h m | |
|---|---|---|---|
| Endeavour Straits . . . . . . . | 9 25 | 1 30 | Cook. |
| Liverpool River and Goulburn Island . . . . . | | 6 0 | (King, ii. 309.) |
| Alligator River, in Van Diemen's Gulf ⎱ (a gulf with narrow entrance) ⎰ | 8 48 | 8 15 | |
| Port Cockburn . . . . . . . . | 8 42 | | |
| St. Asaph's Bay . . . . . . . . . . | | 5 45 | (ii. 237.) |
| King's Cove . . . . . . . . . . . | | 5 15 | (ibid.) |
| Vansittart Bay . . . . . . . | 8 22 | 9 15 | (ii. 324). |
| Montague Sound . . . . . . . . | | 12 0 | (ibid.) |
| Careening Bay . . . . . . . | 8 20 | 12 0 | (ibid.) |
| Prince Regent's River . . . . . . . . | | 12 0 | (ibid.) |
| Roebuck Bay . . . . . . . . | 8 8 | 30 feet, | King, Dampier. |

At Roebuck Bay the tides flow from the westward (KING, ii. 355); so that probably the progression would no longer go on. I find, however, no further observations on the tides in Capt. KING's Survey, though the land extends to $7^h 32^m$ long. E.

The time at Careening Bay shows that we have there the cotidal line of $3^h 40^m$

Returning to the east coast of Australia, and proceeding southwards from latitude 35°, we find the times to become later, so that there is *a point of divergence* on the eastern coast.

The following tide comes from the east (FLINDERS, &c., Plate VII.).

| | Lat. S. ° ′ | Tide-hour. h m |
|---|---|---|
| Bay at the entrance of Banks's Strait . . | 40 45 | 9 0 |

After this the shore runs to the west, and we have the following tides :

| | Long. E.<br>h  m | Tide-hour.<br>h  m | |
|---|---|---|---|
| Corner Inlet (with a narrow entrance) | 9  45 | 11 15 | Bass, in Flinders's Chart. |
| Port Dalrymple (north side of Van Diemen's Land) | 9  47 | 11 45 | Flinders's Chart. |
| Head of Spencer's Gulf (a deep inlet) | 9  11 | 2 15 | ——— |
| Nepean Bay, in Kangaroo Island (with a narrow passage to the east) | 9  11 | 4  0 | ——— |
| Thorny Passage | 9  4 | 12  0 | ——— |

The first and the last of these places may be looked upon as fitted to give the position of the cotidal lines on the coasts; the others are probably too much affected by the retardation of the inlet. Therefore the 3$^h$ cotidal line passes near the point of the coast which is in 9$^h$ longitude east.

I do not find any information concerning the tides of any part of Australia to the west of this. From the comparative narrowness of the passage to the north, it is almost certain that these tides must come from the southern side of the continent; and we have found reason, as already stated, for believing that these southern tides extend to a portion of the N.W. coast.

## On the Tides of the Indian Ocean.

If we examine Mr. Lubbock's Chart, it will appear that the cotidal lines of 11$^h$, 12$^h$, 1$^h$, (Greenwich time,) run along the skirts of the great open space of the Indian Ocean. Thus we have on the south coast of Sumatra,

| | Long. E.<br>h  m | Tide-hour.<br>h  m | Green. Time.<br>h  m | |
|---|---|---|---|---|
| Bencoolen* | 6 50 | 5 50 | 11  0 | |
| Cracatoa Island, Strait of Sunda | 7  2 | 7  0 | 11 58 | Norie. |
| Acheen, west end of Sumatra round the point | 6 22 | 9  0 | 2 38 | ——— |
| Ceylon: Trincomalee | 5 25 | 6  0 | 0 35 | ——— |
| Maldive (King's Island) | 4 52 | 2  0 | 9  8 | Horsburgh, 306. |
| Chagos Island (Solomon's Island) | 4 50 | 1  0 | 9 10 | Norie. |
| | | 1 30 | 8 45 | Lubbock. |
| Mahee Islands (6 feet) | 3 42 | 5 30 | 1 48 | Horsburgh, 126. |
| Amirante Islands (Remire) (9 feet) | 3 36 | 3 30 | 11 54 | ———, 127. |
| African Islands (8 feet) | 3 37 | 9 39 | 6  2 | ———, 127. |

* Norie gives for the time at Fort Marlborough, Bencoolen, 0$^h$ 0$^m$ which does not agree with any of the neighbouring parts.

These three statements of HORSBURGH are so irreconcilable with each other (for the islands are small and near together,) and with the general course of the cotidal lines, that I shall reject them. Mr. LUBBOCK gives from other authority 12$^h$ for the Greenwich time at the Mahee Islands, 12$^h$ 15$^m$ at the Amirante Islands, and 12$^h$ at St. Laurent, a little to the south. This will agree with the general course of the lines.

We have also,

|  | Longitude. h m | H. W. Time. h m | Green. Time. h m |  |
|---|---|---|---|---|
| Roderigue Island (6 feet).............. | 4 13 | 12 30 | 8 17 | NORIE. |
|  |  | 12 45 | 8 32 | HORSBURGH, 114. |
|  |  | 3 13 | 11 0 | LUBBOCK. |
| Mauritius (Port Louis)................ | 3 50 | 12 30 | 8 40 | NORIE. |
| Bourbon ......................... |  | 1 5 | 9 15 | LUBBOCK. |

It will be seen that these data are extremely discordant. A coast so extensive as that of Madagascar may be expected to offer more consistency. We have the following data in NORIE.

East coast of Madagascar:

|  | Longitude. h m | H. W. Time. h m | Green. Time. h m |
|---|---|---|---|
| Samatava Point..................... | 3 18 | 4 18 | 1 0 |
| Fort Dauphin...................... | 3 8 | 4 30 | 1 22 |

West coast of Madagascar:

|  |  |  |  |
|---|---|---|---|
| St. Augustin's Bay ................ | 2 54 | 4 30 | 1 36 |
| Makumba Island .................. | 3 3 | 4 45 | 1 42 |
| Majambo Bay ..................... | 3 8 | 4 30 | 1 22 |
| Minow Island .................... | 3 9 | 5 0 | 1 51 |
| Luza River (entrance) .............. | 3 10 | 4 30 | 1 20 |
| Passandava Bay.................... | 3 13 | 5 0 | 1 47 |

Mozambique Channel:

|  |  |  |  |
|---|---|---|---|
| Johanna Island .................... | 2 57 | 3 0 | 0 3 |
| Sofala ........................... | 2 19 | 4 0 |  |
| Great Comoro ..................... |  |  | 2 30 LUBBOCK. |

We have also in the neighbourhood of the Cape of Good Hope,

|  | Long. E. h m | H. W. Time. h m | Green. Time. h m |
|---|---|---|---|
| Plettenburg Bay ................... | 1 34 | 3 10 | 1 36 |
| Algoa Bay........................ | 1 42 | 3 20 | 1 38 |

It is obvious that the tide is at nearly the same time on the whole coast of the island, but a little later on the inner side; and as we cannot suppose these determinations to be all widely erroneous, we may suppose that the 1$^h$ cotidal line passes near the eastern coast of Madagascar.

The time at Zanzibar on the coast of Africa (E. long. $2^h 37^m$) is $4^h 45^m$, which gives Greenwich time $2^h 8^m$. Mr. Lubbock has $1^h 15^m$. The lines may be accommodated to either, or to the mean*.

I shall, upon the preceding data, suppose the $12^h$ cotidal line to touch the coasts of Sumatra and of Ceylon, to descend southwards near the Mahees and between Mauritius and Madagascar, and to reach perhaps the meridian of the Cape of Good Hope. The $1^h$ line will of course be outside this, and will be inflected and broken by the various bays and islands which occupy the skirts of the Indian Ocean. Within the $12^h$ line, the $11^h$, $10^h$, $9^h$ run, in some measure following its course; the $9^h$ through the Maldives and Chagos Islands. Within this again will fall the lines of $8^h$, $7^h$, and $6^h$. It would appear that the latter must pass near the islands of St. Paul and Amsterdam, and Kerguelen's Land; for we find these statements:

|  | Long. E. | H. W. Time. | Green. Time. |
|---|---|---|---|
|  | h  m | h  m | h  m |
| St. Paul | 5  9 | 11  0 | 5  51  Horsburgh, 83. |
| Christmas Harbour, (Kerguelen's Land) | 4  36 | 10  0 | 5  24  Norie. |

The $1^h$, $2^h$, $3^h$, &c. cotidal lines near the south shore of New Holland are, as we have seen, much retarded by the shore. Hence at a distance from the coast they will bend forwards, and finally northwards and eastwards, so as to meet the wave which comes more slowly still through Torres Straits. From these considerations we may complete the course of these lines with some approximate probability.

In following the cotidal lines into narrower seas, as the Bay of Bengal and the Arabian Sea, they become closer, according to the general rule. And, according to Mr. Lubbock's chart, the tide which arrives at Madras at $4^h 15^m$ Greenwich time, is at Calcutta at $5^h$, and reaches the coast of Aracan at $7^h$.

It is difficult to reconcile Norie's statements concerning the Nicobar and Andaman Islands in the Bay of Bengal. He gives for the former $9^h 15^m$, and for the latter, which are $5°$ to the north of the others, $4^h 30^m$.

---

* Mr. Horsburgh gives the following tide times in this neighbourhood.

P. 180. Mongallon .............................. $3^h 45^m$
    181. Quiloa .................................. 3  45
    183. Zanzibar .............................. 4  30
         *Mombas* ................................. 12  0
    184. Patta ................................... 4  30

In the Arabian Sea the tides also come from the south.  The wave which is on the coast of Africa at the equator at $2^h$ Greenwich time, reaches the shore of Arabia some hours afterwards.   We have,

| | Latitude. | H. W. Time. | Green. Time. | |
|---|---|---|---|---|
| | ° | h  m | h  m | |
| Socotora Island (at the entrance of the Sea of | | | | |
| Bab-el-Mandel)........................... | 12  22 N. | | | |
| Cape Morebat ........................... | 17   0 | 9   0 | 3  40 | HORSBURGH, 220. |
| Mazeira Island ........................... | 20   0 | 10  48 | 5   9 | ————, 221. |
| Ras al Gal (at the entrance of the Persian Gulf) | 22  22 | | | |

On the Indian side of this sea at Kurachee, Kutch, Bombay, and Goa, the tide arrives nearly at the same time; that is, at $11^h$, $11^h$ $30^m$, and $12^h$ after noon, and at about $6^h$ $30^m$ Greenwich time; so that the tide wave must be here nearly parallel to the shore.

With such materials as we at present have, it would be useless to attempt to trace the tide further into the narrower seas, as the Red Sea and the Persian Gulf.

Many maps and charts of various portions of the coasts which border the Indian seas have been constructed from surveys made by direction of the East India Company.   I have examined a collection of these, but have found very few notices of tides, and scarcely any for those parts which would be most decisive with regard to the form of our cotidal lines; as the south coasts of Ceylon, Sumatra, Java, and the detached islands and archipelagos.

## On the Tides of Rivers.

When the tide-undulation reaches the mouth of a river, it enters in the manner which we have described for any branch tide.  The line of the wave revolves round the promontory which it first reaches, of the two which form the opening of the river, and in this way soon meets the opposite promontory. After this it assumes a direction at right angles to the course of the river, and advances regularly up the stream.   There are some circumstances which require notice in such tides.

## Magnitude of River Tides.

If we trace the tide-hours from the sea on one side up the seaward part of the river, and down the other side to the sea again, we may consider the banks as

2 D 2

a portion of the shore in which there is a point of convergence of tides; and the magnitude of the tides will be increased by this convergence, as happens in cases of convergence in general. This augmentation may amount to something very considerable, as in the case of the tides of the Wye at Chepstow, which rise sixty feet. If we consider these very high tides of convergence as a mechanical question, they may be accounted for by what is called "the principle of the conservation of force." When any quantity of matter is in motion, its motion is capable of carrying every particle of the mass to the height from which it must have fallen to acquire its velocity; but if the motion be employed in raising a smaller quantity of matter, it is capable of raising it to a height proportionally greater. In bays and channels which narrow considerably, the quantity of water raised in the narrow part is less than in the wider, and thus the rise in such cases is greater.

When the tide has reached the part of the river beyond which the width does not greatly vary, the wave advances up the river, but its height gradually diminishes by the various causes (friction, impact against obstacles, &c.) which absorb its "force." The great tide of the Wye at Chepstow is, in this way, extinguished before it reaches Tintern.

We may trace this extinction of the tide in the Thames by means of observations which Mr. RENNIE has had made at various points. The following are means of two groups of four days in different parts of a lunation in January 1832.

Height of the tide at

| | ft. | in. | | ft. | in. |
|---|---|---|---|---|---|
| London Dock | 18 | 0 | | | |
| Putney | 10 | 2 | | 8 | 0 |
| Kew | 7 | 1 | | 4 | 8 |
| Richmond | 3 | 10 | | 1 | 10 |
| Teddington | 1 | $4\frac{1}{2}$ | | 0 | $8\frac{1}{2}$ |

In like manner in the Gironde we have first an increase of height by contraction, and then a diminution by the usual impediments. The following heights succeed each other in the course of this river. (DAUSSY, p. 80.)

| | |
|---|---|
| Cordovan | 7·28 feet (French). |
| Saint-Surin ($3\frac{1}{2}$ leagues within) | 7·62 |

La Marechale (7 leagues within)   .   .   8·21 feet (French).
Isle of Patiras (opposite Blaye)   .   .   .   8·41
Bordeaux   .   .   .   .   .   .   .   .   .   .   7·36

In the River of Amazons the tide is still sensible at two hundred leagues from the mouth of the river. (LALANDE, p. 153.)   It occupies several days in ascending to this point; and it has been calculated that there are eight tides travelling along the river at the same time at proper intervals.

In the River St. Lawrence the tides penetrate 432 miles up the main channel, to a point between Montreal and Quebec. (STUART's America, i. 162.)

### The Bore in Tide Rivers.

When the tide is made to rise greatly by the contraction of its channel in the manner just spoken of, the part of the water so affected may be abruptly terminated on the inland side, the depth and quantity of the water on that side not allowing the surface there to be immediately raised by means of transmitted pressure.   A tide-wave thus rendered abrupt has a close analogy with the waves which curl over and break on a shelving shore.   Such a tide-wave is called a *bore*, and in many places occurs in considerable magnitude; producing great noise by the large amount of intestine motion of its particles, and appearing to travel with great rapidity, though it in fact moves slower than the tide-wave under any other circumstances.

The *bore* which enters the Severn is nine feet high; that in the Creek of Fundy is said to be still higher.   In the Garonne this phenomenon takes place near Bordeaux, and is called there *le Mascaret*; at Cayenne it is called *Barre*; it occurs in the River of Amazons, at the junction of the Arawary, having there a face twelve or fifteen feet high, and producing a noise which may be heard at two leagues distance; it is called by the Indians of the neighbourhood *Pororoca*.   (LALANDE, Art. 202.)

### High and Low Water in Rivers.

In the sea, the times from high to low water, and from low to high water, are nearly equal; but in rivers these two intervals may be very different.   It has appeared that tides *die* in travelling up a river; and it will be easily conceived that a small tide would travel a shorter distance than a large one before its extinction. Therefore, a rise of tide which occurs in the lower part of a river must have a certain magnitude, in order that it may affect a point in the upper part *at all*; and the effect on this upper point will be only that which arises from the excess of the tide at the lower point above the amount which is just perceptible above.

Hence the tide at the upper point will not rise during a time corresponding to the whole rise below, but only during a certain portion of the time of highest water; and if we call the state of the river before it begins to rise, *low water*, the time from low water to high water will be less than six hours, and may be much less. This interval also, as follows from the above reasoning, will diminish in ascending the river.

Mr. RENNIE's observations give us the means of verifying the statement just made. It appears that the interval from low water to high water thus taken, is about four hours at Putney, three hours at Kew Bridge, two hours at Richmond Bridge, and one hour and three quarters at Teddington.

In channels open at both ends, the time of high water and the time of slack water are generally different, as has already been explained; but in a river, if the tide-current ever do overcome the river stream, so as to make the water run up, the time of highest water will be very near the time of slack water. At Richmond these times are not found to differ more than a few minutes. It is to be observed, however, that there may be tides in a river without any up-current. The diminution of the velocity of any part will make the water accumulate behind that part; and we may have the tides rise to any height, though the water is flowing down all the while. We have already described such a case.

In open seas, the same kind of curve forms the surface at all periods of the tide, and its convex parts are carried from one place to another. But in rivers, the form and condition of the surface are quite different at high and at low water; and the tide elevation, which is carried from one place to another, has at every place a different relation to the average surface.

We may consider the tides in the Thames, for instance, to take place in the following manner.

When not affected by the tide, the surface of the water slopes so that, compared with Trinity high water mark, it is below it by the following quantities:

|  |  | ft. | in. |
|---|---|---|---|
| At Teddington . . . . . . . . . . | | 1 | 2 |
| — Richmond . . . . . . . . . . | | 4 | 9 |
| — Kew . . . . , . . . . . . | | 8 | 6 |
| — Putney . . . . . . . . . . | | 12 | 0 |
| — Old Swan . . . . . . . . . . | | 15 | 3 |
| Fall at London Bridge (the old bridge). | | | |
| — Billingsgate . . . . . . . . . | | 19 | 7 |

But when it is high water, the whole surface becomes much more nearly level; for we have, taking the mean of the same two groups of four days, which we used before;

Declivity at high water,

|  | ft. | in. | ft. | in. |
|---|---|---|---|---|
| From Teddington to Richmond Bridge . . | 1 | $5\frac{3}{8}$ . | . 2 | $0\frac{3}{4}$ |
| —— Richmond Bridge to Kew Bridge . . | 0 | $5\frac{5}{8}$ . | . 0 | $9\frac{1}{4}$ |
| —— Kew Bridge to Putney Bridge . . . | 0 | 7 . | . 0 | $9\frac{1}{2}$ |

These high waters are not exactly contemporaneous; but this circumstance will not affect the general facts. It appears therefore that the tide raises the surface of the lower part of the river, so as to diminish the slope of that part; that this altered surface extends successively higher and higher up the river, and is nearly horizontal at high water; after that, the tide-surface again descends, and the effects of it disappear in succession from the parts of the river, beginning with the upper parts, till the low water surface is again attained.

### Velocity of the Tide-wave in Rivers.

The velocity of the tide-wave in these as in other cases must be estimated by the rate at which the points of high water are transferred.

### Tides of the River Thames.

Taking the mean of the same days as before, I find time of high water travelling

From London Bridge* to Putney (distance, $7\frac{3}{8}$ miles) . . $0^h 31^m$ . $36^m$
—— Putney to Kew (distance, $5\frac{3}{4}$ miles) . . . . . . $0\ 17\frac{1}{2}$ . $23\frac{3}{4}$
—— Kew to Richmond (distance, 3 miles) . . . . . $0\ 22\frac{1}{2}$ . $26\frac{1}{4}$
—— Richmond to Teddington (distance, $2\frac{3}{4}$ miles) . . $1\ 18\frac{3}{4}$ . $37\frac{1}{2}$

It is clearly seen here that the velocity of the wave diminishes as we ascend the river.

The whole distance from London Bridge to Richmond is sixteen miles, and it appears that the mean of the whole times employed is $1^h 18^m$, which gives an average velocity of about twelve miles an hour.

It will follow from what has already been said, that the point of low water travels up the river with a velocity different from the point of high water. Taking a group of four days, I have the following mean time of *low water* travelling

From London Bridge to Putney.
—— Putney to Kew . . . . . . . $1^h\ 3\frac{3}{4}^m$
—— Kew to Richmond . . . . . $1\ 30$
—— Richmond to Teddington . . . $1\ 30$

I add the following statements of the motion of the tide-wave in rivers.

### Tides of the Bristol Channel and its Rivers.

The tides of the Bristol Channel may be considered as river tides. The following is the progress of the tide-wave, as determined principally by the survey of Lieut. DENHAM.

|  | H. W. Time. | Height of High Tides. |
|---|---|---|
| Lundy Island . . . . . . . | $5^h\ 15^m$ . . . | 27 feet. |
| Hartland . . . . . . . . | 5  20 . . . | 26 |
| Appledore . . . . . . . . | 5  30 | |
| *Barnstaple Reach* (6 miles up) . | 6   0 | |
| *Bideford Branch* . . . . . . | 6   0 | |

* The time at London Bridge is here found by adding $10^m$ to the observed time at the London Docks.

| | H. W. Time. | Height of High Tides. |
|---|---|---|
| Ilfracombe . . . . . . . | 5ʰ 45ᵐ | |
| Fairland . . . . . . . | 6  10 | . . . 33 feet. |
| Minehead . . . . . . . | 6  30 | . . . 38 |
| Sully . . . . . . . . | 6  45* | . . . 31 |
| King's Road . . . . . . . | 6  45 (NORIE) | 46 |
| Bristol . . . . . . . . | 7  0 —— | . 40 |
| Chepstow . . . . . . . | 7  30 —— | . 70 |

From Hartland Point to King's Road is about ninety miles; it appears that the tide travels over this distance in about an hour and a half. The tide increases as it advances.

*Tides of the Frith of Forth and its River.* (STEVENSON's Bell Rock Lighthouse, p. 80.)

Time of high water at

| | |
|---|---|
| Bell Rock . . . . . . . . . . | 1ʰ 30ᵐ |
| Can Rock . . . . . . . . . | 1  45 |
| Elie . . . . . . . . . . . | 2  0 |
| Kinghornness . . . . . . . . | 2  15 |
| Queensferry . . . . . . . . . | 2  45 |
| Alloa . . . . . . . . . . | 3  45 |

Distance of Alloa from the Bell Rock, about seventy miles.

*Tides of the River Waveney (Suffolk).*

This river enters the sea at Yarmouth, and communicates by small and tortuous channels with the Lake Lothing near Lowestoff, which lake is separated from the sea only by a narrow strip of sand-hills. The distance of this communication is not above fifteen miles; but the channels were so small, that originally the tide occupied about six hours in reaching the lake; consequently, it was high water in the lake when it was low water in the sea on the other side of the bank, and *vice versâ*.

The canal lately made in the formation of the Lowestoff navigation has substituted a shorter and straighter course for the former communication,

* From a fortnight's observations of the Rev. W. D. CONYBEARE :—communicated by him.

and the time of high water at the head of the channels is rendered earlier by an hour or two.  Also, a communication has now been cut between the lake and the sea, and flood-gates placed so as to open either way, as the state of the tide may require.

### *Tides of the River Gironde.*   (DAUSSY.)

Times of high water:

| | H. W. Time. |
|---|---|
| Cordovan . . . . . . . . . . | $3^h$ $59^m$ |
| Saint-Surin . . . . . . . . . | 4  33 |
| La Marechale . . . . . . . . | 5   0 |
| Patiras (Isle) . . . . . . . . | 5  21 |
| Blaye . . . . . . . . . . . | 5  35 |
| Bordeaux . . . . . . . . . . | 6  54 |

Distance from Cordovan to Bordeaux, about fifty miles.

### *Tides of the River Elbe.*

| | H. W. Time. |
|---|---|
| Heligoland . . . . . . . . . | $11^h$ $0^m$ |
| Red Buoy . . . . . . . . . | 12  0 |
| Cuxhaven . . . . . . . . . | 1  0 |
| Hamburgh . . . . . . . . . | 6  0 (NORIE.) |

Distance, from Cuxhaven to Hamburgh, about sixty miles.

### *Tides of the River Weser.*

| | H. W. Time. |
|---|---|
| Island of Wrangeroog . . . . . . | $12^h$ $0^m$ |
| Bremen . . . . . . . . . . . | 6  0 (NORIE.) |

Distance, sixty miles.

### *Tides of the Gulf and River of St. Lawrence.*

The tides of the Gulf and River of St. Lawrence have been observed in several places in the course of the survey of those coasts executed by Captain BAYFIELD and Lieutenant COLLINS.

The tide-wave enters the Gulf between Cape Ray (the south point of New-foundland,) and Cape Breton, where the tide-hour (Greenwich time) is about $1^h$. It proceeds west to the Isle of Anticosti, at the east end of which there is a point of divergence, and a point of convergence at the west end.   We have the following data, in the time of the respective places:

| | Long. E. of Quebec. | H. W. Time. | Height. |
|---|---|---|---|
| *Mingan* (N. shore) | 8° 0′ | 2ʰ 20ᵐ | 9 feet. |
| Mahane | 3 40 | 2 0 | |
| Bic | 2 21 | 2 15 | 14 |
| Bersemis | 2˙30 | 2 15 | |
| Goreen Island | 2 50 | 2 15 | 16 |
| Tadousac | | 3 0 | |
| Hare Island | | 3 15 | |
| Kamourasca | 2 20 | 3 45 | |
| Coudres | | 4 40 | 17 |
| St. Thomas | | 5 15 | 18 |
| Isle of Orleans | | 5 45 | 18 |
| Quebec | | 6 45 | 20 |

The longitude of Quebec is 71° 10′, or in time 4ʰ 45ᵐ.   Therefore the tide-hour, Greenwich time, at Mahane is 6ʰ 30ᵐ nearly, and at Quebec 11ʰ 30ᵐ.

We may observe here, as in the Bristol Channel and in the Garonne, the increase of height in the tide as it advances.

The distance from Mahane to Quebec along the river is about three hundred and fifty miles.   The tide-wave occupies, as we see, about 5 hours in travelling over this space.

### Sect. IV.  *General Remarks on the Course of the Tides.*

#### 1. *On the Velocity of the Tide-wave.*

The ridge of water which brings the tide moves from the position of one of the horary cotidal lines to that of the next, in an hour; hence the velocity with which this wave advances will be measured by the distance of two such lines, this distance being taken in a direction nearly perpendicular to both. It will be seen by a glance at the map, that the velocity so measured is very different in different places.   In latitude 60° S., where the sea is not interrupted by any land, except the narrow promontory of Patagonia, nearly the whole circumference of the globe must be occupied by twenty-four cotidal

2 e 2

lines; hence the velocity in an hour will be about 670 miles. In the Atlantic Ocean, the distances of the lines appear to be in some cases 10° of latitude, or near 700 miles. This wave travels from the south point of Ireland to the north point of Scotland in about eight hours; this distance is about 7°, or 160 miles an hour along the shore. On the eastern coast the velocity is less; from Buchanness to Sunderland, three hours, gives about sixty miles an hour; from Scarborough to Cromer, also three hours, about thirty-five miles; from the North Foreland to London in two hours, makes the speed about thirty miles an hour; from London to Richmond in one hour and a quarter, gives us thirteen miles in that part of the river.

The velocity which results from theoretical considerations is not easily determined. NEWTON's solution of the problem of waves proceeds upon a supposition of the motion of the fluid so different from the real case, that it has no claim to be considered even an approximation. If we suppose any accumulation of fluid in one part of an ocean, this accumulation would transmit its pressure in all directions; and by the play of such pressures in different parts of the fluid, motions of the parts of the fluid would be produced, which would transfer the elevation of the surface to other situations, as really takes place. But it is very difficult to deduce the results of this action, without additional suppositions. One of the simplest suppositions is that on which LAGRANGE gave the solution; namely, that the horizontal motion of the parts of the fluid is the same all the way to the bottom. On this hypothesis, the velocity with which a wave is transmitted along a uniform channel, is that which a heavy body would acquire by falling freely through half the depth of the fluid. On other suppositions, the velocity of transmission would depend on the dimensions of the wave.

The velocity of waves has not been very completely determined by experiment. In WEBER's experiments (Wellenlehre, p. 172), the velocity in a channel six inches deep was nearly that which LAGRANGE's theory would give, but the velocity did not increase so fast by increasing the depth, as the theory would require. Perhaps we might represent the results by supposing that in deeper channels a portion of the *depth* is not *effective*, the water at the bottom not being affected by the horizontal motion.

Nor have we any good observations of the velocity of waves at sea. A persuasion appears to obtain that the velocity depends on the breadth of the wave, and that the larger waves travel fastest.

The tide-*wave* would be transmitted by the same laws of the effects of pressure in fluids, which produce the transmission of other waves, though the tide-wave is so different from others in its dimensions, being very broad and flat. On the east coast of England, for instance, the breadth of this wave reaches from the Orkneys to the coast of Norfolk, which are the places where its two ridges contemporaneously are,—the hollow of the wave being then on the coast of Yorkshire, and perhaps twelve or fourteen feet lower than the ridges. And in the Atlantic the wave is still wider, one ridge being in the Antarctic Ocean, while the other is at Newfoundland.

It will be seen from what has been said, that we cannot trace very distinctly the causes which produce so great differences in the velocity of the tide-wave in different places. The difference of depth has probably a great share in producing these differences; and accordingly we find that in narrow seas, which are also generally the shallowest parts of the ocean, the velocity is much diminished, and the tide-waves consequently *thronged:* and this fact appears with remarkable uniformity in every part of the map. Other causes of the difference of velocity are probably the obstacles to motion offered by the uneven form of the bottom and shores: and as we are in a great measure ignorant of the form of the bottom of the sea, we may easily conceive that the course of the cotidal lines may exhibit many inexplicable anomalies.

If we calculate the *effective depth* of different parts of the fluid on LAGRANGE's principle, from the velocities above stated, we find that the effective depth of the Thames from London to Richmond is 13 feet; from the Nore to London, 90 feet: the effective depth of the sea on the east coast of England, 120 feet; of Scotland, 360; of the Atlantic, west of Ireland, 2600 feet; of the Atlantic in its middle parts, 50,000 feet, or above nine miles.

## 2. *On the Form of the Cotidal Lines.*

It will be seen in the map that the form of the cotidal lines is very various, and apparently anomalous. Still they conform in most of their general circumstances to the principles which were stated at the outset. They are convex in the direction of their motion, the ends near the shore being held back by the smaller velocity in shallower water and other resistances, agreeably to what has just been stated : they bend round promontories and islands, forming a point of divergence on the side of the coming tide ; tides which meet two distant points of the same shore at the same or nearly the same time, produce a point of convergence between them ; and other features appear, such as the theory requires.

It may, perhaps, appear to some persons that the forms of the cotidal lines, as I have given them, are in some cases too irregular and complex to be probable; as for instance, on the west coast of America.   To this I can only say, that they are the simplest forms, and the most agreeable to the general rules of the rest, by which I can connect such observations as we possess.   If it shall appear that our observations are inaccurate, the lines may very probably be in reality simpler than they are here represented : and if this representation be at all correct, it ought to be confirmed by observations made at many adjacent points on the coasts which exhibit these complex forms.

There is one inaccuracy in the position of these lines which I have not attempted to remove, as it does not materially alter their shape.   The times of high water from which the lines are constructed, are the times *at* the full and change of the moon ; but the times for which they ought to be constructed, are the times of the tides *corresponding to* the full and change, which tides may be one, two, or more days after the syzygy.   Thus, the time which we have used for Plymouth is $5^h$ $33^m$, that being the time of high water at full and change, or the *vulgar* establishment.   But the high water at Plymouth corresponding to the moon's syzygy is two days and a half after the change, at which time the tide is only $5^h$ $3^m$ after the moon, this being the *corrected* establishment. This cause, however, will not alter the form of the line ; for it applies equally to the tide at every point of the same line: the only consequence is, that the line which is taken to belong to $5^h$ $33^m$, does in fact belong to $5^h$ $3^m$, *corrected* establishment; and the $5^h$ $33^m$ line, similarly corrected, would be a little in advance of the position we have assigned ; and each of the other lines will in like manner require to be thrown forwards by a small quantity proportional to the age of the tide.   When the course of the lines is obtained with tolerable accuracy, this can easily be done.

### 3. *On Tide-Currents.*

The motions of the currents produced by the tides are, in most circumstances, much more important and interesting to the sailor than the motion of the co-tidal lines: but their laws are far more complex and difficult than the laws of the transmission of high water, nor shall I at present attempt to treat of them in detail.   I will observe, however, that complex as the problem is, I do not think it would be a desperate undertaking to assign some of the leading rules of the tide-currents, if the rules of high water were accurately and extensively ascertained; a good approximation to the solution of the problem on which we are now engaged must precede anything like a scientific view of the phenomena of tide-currents.   I will only offer a few remarks.

One of the most important circumstances of this subject is that which has been already stated,— that in an open channel the *flood current* (the current which runs till high water) will continue running for three hours afterwards, or till *half-ebb*; and the ebb current, which then begins, will run after low water till half-flood.   The time of slack water is intermediate between the times of high and low water.

In proportion as the channel is obstructed at the further end, the flood current runs for a shorter time after the flood; and in a closed creek the flood current ends at high water.

Thus in the British Channel the flood current runs three hours after high water; in the Bristol Channel, two hours after high water; in rivers, only a few minutes after.

### *Revolving Tide-Currents.*

We have, in what precedes, spoken only of currents which run alternately in two opposite directions; but in many seas the currents at successive periods shift into many new directions.   In particular, they in many places revolve in twelve hours through the whole circuit of the compass.   This is noticed with regard to the tide-currents among the Scilly Islands by SPENCE (p. 4.), and has been particularly attended to by Captain M. WHITE and other officers who have surveyed our shores.   The Dutch have noticed this phenomenon on their own coasts; and have expressed it on their charts by a peculiar mode of representation.   It has been expressed in a different manner by Captain THOMAS in a

Chart of the German Ocean: we may hereafter say a few words on the selection of a mode of denoting the facts of such cases. As a specimen of these cases I give the following from Captain WHITE:

Stream of tide off Scilly in the vicinity of the Poul Bank:

| | | |
|---|---|---|
| From low water to half-flood . . . | N. 45° W. velocity | 1 mile per hour. |
| —— half-flood to 4 hours flood . . | N. 22 E. . . . | 1 —————— |
| —— 4 hours flood to high water . . | N. 56 E. . . . | 1·6 —————— |
| —— high water to quarter-ebb . . | S. 67 E. . . . | ·75 —————— |
| —— quarter-ebb to half-ebb . . . | S. 22 E. . . . | ·75 —————— |
| —— half-ebb to three quarters ebb . | S. 22 W. . . | ·5 —————— |
| —— three quarters ebb to low water . | W. . . . . | 1·0 —————— |

It will be seen from this, that here the current goes round from the west by the north to the east, and so to the south; that is, it revolves in the same direction as the sun does.

These revolving currents take place especially in the neighbourhood of places where the streams of tide separate or meet: and it is not difficult to see that these are the situations in which they might be expected to occur. Thus at the Scilly Islands, the Atlantic tide-stream divides into two; one of which runs up the Bristol and St. George's Channels, while the other proceeds along the English Channel. At the point where the stream bifurcates, if the two branch currents, while they are in contact, exactly balance each other in lateral pressure, they may run in constant directions; but considering the various circumstances which influence each of the two, this can hardly be the case for the whole time of their motion; if one of the currents which are thus in contact become more powerful than the other, it overrides and turns it, and thus gives rise to a change of direction; and as the circumstances of each current go through a regular cycle in twelve hours, it may easily be imagined that the direction of the stream may go round the compass in that time.

It would not be to our present purpose to trace much further the details of such cases, but we may observe that they have attracted notice as supposed indications of the direction in which the tide comes. Thus it has been observed, that on the Dutch coast the tides northward of the Oude Steen bank of Goeree follow the course of the sun, while at the entrance of the Schelde and off Flushing they revolve in the opposite direction. (Sailing Directions, North Sea, p. 54.)

We may observe also, that we have frequently near the shore a *counter stream* running in a direction opposite to the main tide stream which prevails further seawards, (for example, the north-east coast of Ireland). This commonly occurs where the stream is impeded in its progress by some projecting shore or island; and the opposite currents which are thus brought in contact often produce remarkable effects, especially where the promontories are of very sudden projection.

Where a promontory does not produce a counter current, the stream which advances along the shore is thrown obliquely into the main stream by the bend of the land, and we have thus various phenomena, such as the Race of Portland and other similar cases.

### 4  On the Magnitude of Tides.

The magnitude of tides is very various, and the causes of the differences are often far from obvious. I shall notice some of them.

I shall not endeavour to determine the magnitude of the tide according to theory. The smallness of the tide in the middle of wide oceans, as in the islands of the Pacific, where it amounts only to two or three feet, is not to be taken as a proof that this would be the amount on a globe entirely covered with water; for the interruption produced by the continents entirely modifies the circumstances of the general tide. This interruption will diminish the original tide; and the derivative tide which enters such oceans from the south-east, is *diffused* over so wide a space that its amount is also greatly reduced.

In the same manner in which the tides are diminished by diffusion, they are augmented by convergence. The greatest shore tides occur near points of convergence; and when there is added to this a contracting bay or river channel, we have tides of extraordinary height. Thus the head of the Bristol Channel, of the Bay of St. Malo, of the Bay of Fundy, have very large tides.

There is one difficulty which it may be well to notice. Since the tides in places near each other are of different heights, the surface of the water cannot be level: "This unequal rise," says Mr. SPENCE, " is in a manner directly contrary to the known law of all fluids, by which they run to the same level at all neighbouring places, and rise to equal heights from the centre of the earth; as, for instance, at Cape Clear, springs rise only eleven feet perpendicular, but at Scilly twenty feet. At Tusker and Wicklow, springs rise only four or five feet, but in the vicinity of St. David's Head thirteen feet. At Brest I understand that springs rise only three or four feet*, but at Falmouth sixteen or seventeen feet. Among the Norman Isles four or five fathoms is not an uncommon rise, but at the very extremity of Portland Bill I have taken a favourable opportunity to observe the greatest rise of a spring, and found it scarce seven feet."

The difficulty thus stated is not without weight; the solution of it appears to be this;—that the water, on a great scale, cannot be considered as at rest, and that therefore the form of the surface is not that of equilibrium. We may perceive that in a stream of water in rapid motion, as in a mill race, there is a *permanent* curvature of a part of the surface, showing that a fluid in which the parts are moving with different velocities does not assume a surface exactly horizontal. There is a certain form of the surface, such, that with it the forces arising from the weight and the velocity of the parts of the fluid balance each other hydrostatically; and this is a surface such as admits of these inequalities of rise of the tide at different places.

So long as the velocities of the different parts retain their amount, the surface is just as necessarily determined, as it would be determined to be horizontal if all the parts were at rest. Hence if, under such circumstances, the level of one part be altered, the level of all the other parts will be altered in a corresponding manner. This will explain why the *times* of high water at neighbouring places are so nearly the same, though the heights are very different. On the coasts of England and France, for example, if the water fall on one side, this depression is transmitted through the fluid to the other side in a very short time; and in nearly the same time, whether the fluid, independently of this transmission, was at rest or in motion. The supposition of its being already in motion allows of a different level on the two sides; and as the motion may be different at different times, the side which is highest at one time may be lowest at another. Thus, combining the hydrostatical effect of the currents with the laws of transmitted undulations, we perceive how the

* This is an error.

form and motion of the cotidal lines may be very regular, while the tides along the same cotidal line are very different in their magnitudes, and form a very irregular series.

In all the cases which have been carefully examined, the time of high water at neighbouring places has been found to vary continuously and slowly in advancing along the sea; it is probable therefore that the statements of tides at very different hours at adjacent places will, in most instances, be found incorrect.

We may observe, however, that the relative level of the water at neighbouring places may change, by a change in the contiguous currents, and this change may be combined with the transmission of the undulation. The relation of contiguous currents may be different at springs and neaps, in consequence of the different supply of water; and from this cause, and from the difference of depth, the interval of times of high water at neighbouring places may be different at springs and at neaps. If this were the case, it would appear, in the observations, by the times of high water, in the course of a semilunation, having different relations to the establishment. If we make a curve, of which the ordinates represent the interval of time by which the tide follows the moon, this curve would be of different form at two such places.

### 5. *On the Constancy of the Cotidal Lines.*

A doubt may very naturally occur to the reader, whether the cotidal lines which I have endeavoured to trace, have a permanent position on the globe, or whether they do not vary from one tide to another, according to the age of the moon, or to other causes of change. They are here drawn for the tides which occur at new and full moon;—would their form be the same at any other period of the lunation? and may they not be affected by accidental causes, as winds and other atmospheric influences? May they not also experience secular changes, and vary from century to century? To these questions we have no means at present of returning answers with any certainty and accuracy. Our knowledge is yet very imperfect and very doubtful, with regard to the data on which the first approximation to the position of the lines must be conducted. When this is obtained, very multiplied and accurate observations may enable us to determine their monthly and other variations, if such exist.

The mode in which such variations will manifest themselves, will be this;—the *laws* of the time of high water will be different at different points. The curve of which we have just spoken, of which the ordinates are the intervals by which the tide follows the moon, will be different for different places of observation.

It appears probable that the monthly variation is not considerable; for, so far as observations have been made, the times of high water at most places appear to follow nearly the same laws. In the case examined by Mr. LUB-BOCK, (page 19 of the present volume,) this resemblance is very striking. Mr. DESSIOU has there constructed the curves of which I have spoken, for Brest, Plymouth, Portsmouth, Sheerness, and London, and their identity is almost complete. But it would be interesting to have observations of the same kind made in distant seas, where the tide arrives by roads altogether different, instead of comparing branches of the same derivative tide, as all these are.

It appears likely, also, that there will be *some* variation owing to this cause; for the tide of many places may be considered as composed of parts which come to the place by different roads, and the relation of these parts will be affected by the changes of a lunation. If we subtract four feet from the height of each of two partial tides which come in different directions, their proportion will be altered, and thus the time of the component tide may be altered also.

As to the atmospheric causes of variation, we may say in like manner, that the partial irregularities of the tides show that there are some such variations, while the general steadiness of the laws of tide phenomena proves that these variations are not considerable.

With respect to the existence of secular changes in the phenomena of the tides, we are still more in the dark. The formation and removal of sand-banks and bars, the silting-up of rivers, and similar causes, may, and on a small scale must, produce such effects. Mr. LUBBOCK has pointed out (Philosophical Transactions, 1831, p. 389,) the probability that high water in the Port of London takes place at present earlier than it did formerly. The tide observations directed by Mr. RENNIE at Putney, Kew, Richmond, and Teddington, will, it may be hoped, discover the changes produced in the tides of the upper part of the Thames by the removal of Old London Bridge. Whether such changes take place on a larger scale it is impossible at present to say.  It is conceivable that the alteration in the depth and form of the bottom of the ocean produced by great marine currents, and by the materials they bring or take away, may affect the velocity, and, by means of this, the form of the cotidal lines which cross the ocean.  The change must be very slow, if it take place at all.

### 6. *On some Peculiarities of the Tides.*

*Difference of the two diurnal Tides.*—It has been remarked in various places, by separate observers, that the evening tide is higher than the morning tide at one part of the year, and lower at another.   I extract below a few notices to this effect*.

These detached statements are confirmed by a more exact examination.  The Plymouth observations show the difference, though to a small amount.   LA-PLACE, from his examination of the earlier series of observations at Brest, found about half a foot for the excess of the evening tide over the morning tide at the summer solstice.   The examination of the later observations at Brest gave the result slightly larger. (Supp. Mec. Cél. liv. xiii. p. 161.)

This peculiarity was explained by NEWTON.  From the vernal to the autumnal equinox the sun has north declination; and as the moon's orbit is never much inclined to the sun's, a line drawn from the earth's centre to the moon would meet the earth's surface, on the side towards the sun, in north latitude.   Now such a line is the axis of the tide spheroid, supposing the tide to be always under the moon; and the tide taking place when the moon is in the meridian.

---

* Philosophical Transactions, vol. iii. p. 814.  Captain STURMY, at Hong-road near Bristol, says. " Concerning our diurnal tides we observe, that from about the latter end of March till the latter end of September, they are about 1 ft. 3 in. higher perpendicularly in the evening than in the morning; that is, if high water happen after the sun is past the meridian, or in the tides betwixt noon and mid-night : but from Michaelmas till our Lady Day we find the contrary, the day-tides being in that higher by 15 in. than the night tides, or the tides between midnight and noon."

Vol. iii. p. 633. COLEPRESS, at Plymouth, says, " Our diurnal tides from about the latter end of March till the latter end of September, are about a foot higher in the evening than in the morning, that is, every tide which happens after 12 in the day before 12 at night; and *vice versâ* the rest of the year."

Asiatic Researches, 1829, P. 1. p. 262. KYD on the Tides of the Hoogly : " There is another local affection of the tides, the cause of which I cannot satisfactorily explain.   In the north-east monsoon [from the end of October to the beginning of March] the night tides are the highest, whilst in the south-west monsoon [from March to October] the day tides are highest."

is higher as the place is nearer to the vertices or points where this axis of the tide spheroid meets the earth's surface. Hence, in this case, the tides which occur on the side of the earth next the sun, or the day tides, would be larger for a place in north latitude than the tides on the opposite side. For a similar reason the night tides would be highest in winter*.

This would be so on the supposition that the tide took place when the moon was on the meridian; but if the tide take place 6 hours after the moon's transit, the pole of the spheroid having still the same declination as the moon, the pole which follows the moon will be north of the equator, from the time when the moon is 6 hours to the west of the sun till the time when she is 6 hours to the east of him; that is, from the time when the tide is at noon, to the time when it is at midnight: after this the moon goes south of the equator, the largest tide is that opposite to her, and this again happens at times which advance successively from noon to midnight.

But if the tide take place 18 hours after the corresponding transit of the moon, it will appear by similar reasoning, that the lower tide will take place from noon to midnight when the sun is north of the equator, and vice versâ.

If the tide be at 30, or at 54 hours after the transit, the phenomena will be the same as when it is at 6 hours after.

Now the last case appears to be that of the phenomena at Brest, Plymouth †, and Bristol; therefore the tide of these places cannot be 18 hours old, and

---

* LAPLACE has objected to this explanation,—that if it were true, the two semi-diurnal tides at Brest, when the moon has her greatest declination, ought to differ in the ratio of 8 to 1, whereas their difference is very small. But it is clear that NEWTON's theory must be applied, as LAPLACE's also requires to be applied, by considering the tides on our coasts as derivative tides from those in the Southern Ocean. When the moon has considerable south declination, it is clear that the superior tide in the Southern Ocean will exceed the inferior tide; *how much* it will exceed it, must depend on the form of the ocean. LAPLACE is not able by his method, any more than NEWTON by his, to calculate the excess à priori.

† There is an anomaly, as yet unexplained, in the Plymouth Tide-Observations. From the course of the cotidal lines it would appear that the Plymouth tides are only about 1 hour 35 minutes after those at Brest; yet it appears by the examination of the observations, that the age of the tide at Brest is about $1\frac{1}{2}$ day, and at Plymouth $2\frac{1}{2}$ days. The amount of the latter is as great as the age of the tide at London; which circumstance it is difficult to account for, as Mr. LUBBOCK has already noticed (see his "Note on the Tides").

It may be observed that the difference between the law of the Plymouth tides (as given by the ob-

must be either 6, or 30, or 54 hours; it is certainly more than the former, therefore its age is probably 30 hours or a little more, which agrees with what we know from other considerations.

In the port of London this difference of day and night tides is not perceived, for the tide there is compounded of two, distant by 12 hours from each other, and therefore of a larger and smaller tide in each case.

We appear at first sight to have an exception to the rule now stated, in the case of New Holland. It is asserted by Captains COOK, FLINDERS, and KING, that the night tides there are *always* greater than the day tides. COOK was there in *August* 1770, FLINDERS in *January* 1802, and Captain King in *January* 1822, and therefore the observation has been made both when the sun was north and when the sun was south of the equator, which appears to contradict the rule of the other cases.

But on examining further, we find that the observations took place on different parts of the coast. Cook's were made on the eastern side, in consequence of the Endeavour getting within the reef: here the tide-hour is 8; and as the high tide occurs 8 hours after the moon's transit *on the other side* of the equator, the age of the tide here must be about 20 hours. Captain FLINDERS made his observations at King George's Sound and the neighbouring parts, where the tide is 7 or 8 hours later than on the western coast, which throws it into the next half-day, and reverses the phenomena of morning and evening. Captain KING confirmed Captain FLINDERS's observations on this part of the coast.

servations,) and those of other places is easily shown, without any peculiar mode of considering them. The greatest and least intervals between the moon's transit and the tide are as follows, (see Mr. LUBBOCK's Paper):

| | Moon's Transit. | Greatest Interval. | Moon's Transit. | Least Interval. | Difference of Intervals. |
|---|---|---|---|---|---|
| | h m | h m | h m | h m | h m |
| Brest ........... | 9 30 | 4 8 | 5 0 | 2 49 | 1 19 |
| Plymouth........ | 10 0 | 5 48 | 6 0 | 4 12 | 1 36 |
| Portsmouth ...... | 9 30 | 12 2 | 5 0 | 10 41 | 1 21 |
| Sheerness........ | 10 0 | 12 54 | 6 0 | 11 25 | 1 29 |
| London.......... | 10 15 | 2 10 | 6 0 | 12 42 | 1 28 |

The difference of the greatest and least intervals ought to be the same in *simple* tides at all places. The above variety of values of this interval *may* be owing to the different manner in which the tides at the different places are compounded.

Probably a good series of tide-observations at the Eddystone would throw light upon this apparent anomaly. It would be easy to suggest the means by which they might be made.

*Single Day Tides.*—In some places there is high water once only in twenty-four hours. The most celebrated of such cases is that of the harbour of Tonquin, described by Mr. DAVENPORT, attempted to be reduced to rule by HALLEY, and explained from theory by NEWTON. The circumstances of that case are as follow:

The tide rises and falls every day during about 12 hours each way. The rise begins every successive day later by about three quarters of an hour, so that in 15 days the time of high water advances from 1 o'clock in the afternoon, for instance, to 12 at midnight; after which it does not advance to 1 in the morning, but falls back 13 hours to 12 o'clock at noon, and so on perpetually. In this way the high water is always in the afternoon during the summer half-year (March to October), and in the forenoon during the remaining half. (Philosophical Transactions, vol. xiv. p. 162.) About the time when the tide time falls back 13 hours, the tides are very small and scarcely perceptible; at the intermediate times they are greatest.

NEWTON's explanation of these phenomena consisted in supposing that the tides at this place are compounded of two tides arriving by different paths, one 6 hours longer than the other. When the moon is in the equator, the morning and evening tides of each component tide are equal, and the tides obliterate each other by interference, which takes place about the equinoxes. At other periods the higher tides of each component daily pair are compounded into a tide which takes place at the intermediate time, that is, once a day; and this time will be after noon or before, according to the time of year, as will appear by the reasoning of last section*.

This explanation is extremely probable, and requires only to be confirmed by observations of the tides of the adjacent parts, which might give the course of the two component tides before they meet.

It has already been mentioned, that at Juan Fernandez the tide is said to run 12 hours each way; and the same is stated to occur in other cases. But the only instances in which we have accurate accounts of such phenomena are two others, which also are found in the Indian Ocean, one on the north and

---

* I refer to NEWTON's explanation rather than LAPLACE's, as more familiar to the common reader. The two explanations appear to me to coincide, both in the forces to which they attribute the fact, and in the laws of the phenomena to which they lead.

one on the south coast of New Holland.   I shall quote the descriptions of these from Captains FLINDERS and KING *.

* *Tides at King George's Sound; south coast of New Holland*, long. 7ʰ 52ᵐ E. FLINDERS, i. p. 71. January 1802.—" According to Lieut. FLINDERS's observations on shore during 16 days, there was only one high water in 24 hours, which always took place between 6 and 12 at night; for after, by gradually becoming later, it had been high water at 12, the next night it took place soon after 6 o'clock, and then happened later by three quarters of an hour each night, as before.   The greatest rise was 3 feet 2 inches, and the least 2 feet 8 inches.

" The accumulation was made in this manner: after low water it rose for several hours, then ceased and became stationary, or perhaps fell back a little.   In a few hours it began to rise again; and in about 12 from the first was high water.

" It was observed by Capt. COOK upon the east coast of this countryᵃ, and since by many others, including myself, that the night tide rose considerably higher than that of the day, which is conformable to our observations in King George's Sound, but with this difference, that in the day we had scarcely any tide at all."

KING, ii. p. 380, 381, January 1822. Oyster Harbour, a branch of King George's Sound.—" During the springs, high water always takes place at night.   The flood tide in the entrance generally ran 16 hours, and ebbed 8 hours.   High water at full and change took place at 10ʰ 10ᵐ at night; but on the bar the rise and fall was very irregular, and a vessel going in should pay great attention to the depth, if her draught is more than 10 feet; for it sometimes rises suddenly 2 feet.   The spring tides take place about the third or fourth day after new or full moon."

*Tides at Wellesley's Islands; north coast of New Holland*, long. 9ʰ 19ᵐ in the Gulf of Carpentaria. FLINDERS, ii. p. 149. November, 1802.—" The tides in the Investigator's Road [the channel between the islands and the main land,] ran N.N.E. and S.S.W. as the channel lies, and their greatest rate at the springs was 1¼ mile per hour; they ran with regularity, but there was only one flood and one ebb in the day."

ii. p. 155. Bountiful Island, near the former.—" From a little past 10 in the morning to 11 at night the tide ran ½ a mile an hour to the S.W., and N.E. during the remainder of the 24 hours; the first, which seemed to be the flood, was only 3 hours after the moon, above 6 hours earlier than in the Investigator's Road: but the time of high water by the shore might be very different; no greater rise than 5 feet was perceived by the lead line."

I add the following notices from PURDY (A. M. p. 76.), which seem to refer to tides of this kind in the West Indies.

*Vera Cruz.*   Only 1 tide in 24 hours: height 2 feet.

*About the Island of St. Bartholomew* the flood at new and full moon runs S.E., and it is then high water at 10ʰ 30ᵐ P.M. while the sun is furthest to the north of the equator; but comes about two hours sooner in the following months, till the sun gets furthest to the south, when it is high water at 10ʰ 30ᵐ A.M., and it runs afterwards in the same proportion back again.   The greatest difference in the ebbing and flowing is 18 inches, but in general only 10 inches.

The following is not obviously of this class.   (Ibid.)

" In *Ponce*, or *Chatham Bay* (S. end of Florida), it runs 3 hours flood, then 3 hours ebb; next 9 hours flood," &c.

---

ᵃ HAWKESWORTH's Voyages, vol. iii. p. 647.

There can be little doubt that these are cases of interfering tides, but observations at several adjacent places would be requisite to determine the paths by which the interfering tides arrive. In the case of Wellesley Islands in the Gulf of Carpentaria, the extremely broken form of the land north of New Holland allows us easily to suppose such different paths for the tide-wave. At King George's Sound on the south, this condition is less obvious. It is not impossible that the tides in this part may be affected by an undulation propagated from the Indian or South Atlantic Oceans by the way of the south pole.

*Double Half-day Tides.*—In some places we have more than two tides a day. Poole in Dorsetshire is an instance of this.

" *Poole Harbour* has an uncommon advantage, namely, that of the tide ebbing and flowing twice in 12 hours. It is low water at about half-past 3 o'clock, then flows regularly 5 hours and 20 minutes, and makes proper high water about 50 minutes after 8 o'clock. It then ebbs $1\frac{1}{2}$ hour, and again flows $1\frac{1}{2}$ hour, and then ebbs till low water.

" The second flood seems to be owing to the peculiar situation of the entrance; for by its being a bay towards the east, the tide of ebb from between the Isle of Wight and the main falls into that bay, and forces its way into the river, so as to raise the water for an hour and a half; at which period the water without the bar, by its falling below the level of that within, produces a second ebb for more than 3 hours, or till it is low water.

" In *Christchurch Harbour* the tides are nearly similar." (DESSIOU, Sailing Directions, E. Channel, p. 83.)

This explanation is probably right in principle: the level is altered by the velocity of the ebb current near the shore; and this alteration of level, from the hydrostatical effect of currents, shows itself in the form of a second rise of the surface, after it has begun to descend from the true high water.

*Weymouth Harbour* is circumstanced nearly as Poole, and has also a double tide. " On the days of new and full moon it is high water on the shore in Weymouth Harbour about $6^h 30^m$; low water the first time about $11^h$, and low water the second time about $2^h$; so that with spring tides there is about $4\frac{1}{2}$ hours between high water and the first low water, during which time the whole ebb falls; the tide then flows about $1\frac{1}{4}$ hour, and rises a few inches, and again ebbs for about 2 hours and falls a few inches, but is not so low as the first low water.   Neap tides sometimes remain stationary about low water for 2, 3, and sometimes 4 hours.   From which it appears that there is only one high water with springs, but two low waters; and with neaps there are not two low waters observable, but a stationary tide.

" The tides in West Lulworth Cove are much the same as at Weymouth."— MACKENZIE's Survey of the Channel (Admiralty MSS.).

## Sect. V. *Suggestions for future Tide-Observations.*

It has appeared in the course of the preceding discussion, how extremely imperfect, and in many cases contradictory, are the statements which we at present possess concerning the establishments or tide-hours at different places. This has arisen in a great measure from the circumstance, that these observations were made without any settled rule or any definite object.   If, with the opportunities which now exist, observations are for the future made with due attention to the circumstances of real importance, we may in a very few years be able to draw a map of cotidal lines with certainty and accuracy; and thus to give, upon a single sheet, a tide table for all parts of the earth.

## 1. *Of the Observation of the Height and Time of Tides.*

It is perhaps not necessary to go into any detail on the subject of the arrangements which may be useful in observing the height and time of each tide. Any person who gives his attention to the subject, and especially any naval person, will probably hit upon contrivances better adapted to the circumstances of his particular case, than could be pointed out by any general suggestions.   It may, however, be not superfluous to mention one or two resources for the consideration of observers.

The instant to be observed is that when the surface of the water is highest. If the water be perfectly still, the surface changes very slowly when near the highest point, and appears to be stationary for some moments.   To avoid the difficulty produced by this circumstance, some observers have observed, not the time when the water is highest, but two instants of equal height before and after the greatest; and the time of greatest height is supposed to bisect this interval.

In comparative observations, if the time observed in one case be that at which the surface begins to fall, it ought to be so in all cases; and similarly, if the time observed be that at which the surface ceases to rise.

*To obviate the effect of waves* in rendering the surface uncertain, the following apparatus may be used. Let a spout or pipe be fixed upright, in such a situation that at tide time the water reaches its lower part. The bottom of the pipe must be stopped, and a number of small holes (for instance, each half an inch in diameter) must be made in or near the bottom. A float nearly filling the pipe is to be placed in it, and to carry a light upright rod, divided into feet and inches, which are to be read off by means of an index or mark fastened to the top of the tube. The apertures in the bottom of the tube will allow the float to rise and fall with the general surface, without any sensible loss of time; while the smallness of these apertures will prevent the oscillations of the waves from affecting the inside of the tube. The moment when the rod, and consequently the surface, is highest, may be observed by means of the index.

The spout might be further protected, by having its lower end in an open box. It might be fixed to a post driven into the ground. If the most convenient place for it were at a distance from the observer, it might be observed by means of a telescope. If it were desirable, it would not be difficult to devise means by which the rod should pull a trigger and ring a bell the moment it began to descend; and thus mark the moment of high water without the trouble of watching.

Mr. MITCHELL'S tide-gauge, which is in operation at Sheerness, marks the time of high water by means of a curve, of which the abscissas represent times, and the ordinates the corresponding heights. The greatest height during each tide is picked out by the eye, from the curve drawn by the instrument, and the corresponding time taken from the scale. This instrument is found to answer extremely well in practice, but could not be used except in stationary observations.

The time used in tide-observations may be mean or apparent time, but it should always be noticed which is employed, and by what means obtained.

## 2. On finding the Establishment of any Place by Observation.

*The vulgar establishment* of any place, or the time of high water at the full and change of the moon, may be determined roughly by an observation of the tide on the day of the full or change.   We must, however, observe the following corrections.

1. The establishment should be expressed by saying, that the tide is, on the day of full and change, so many hours *after the moon's transit,*—not that it is *at* such an hour *of the day*.   The mode of stating the result here recommended has been recently commonly employed by some of our best naval surveyors; for instance, Captain KING.   The time of the moon's transit is of course easily known from the tables.

2. If the tide be observed according to mean time, and the time of the moon's transit be determined according to apparent time, it will be necessary to apply the equation of time to the interval.

3. The establishment of any place may be determined by observations not made at the full or change, by applying to the observed interval of the tide and moon's transit, a correction depending on the moon's distance from the sun.

This correction is by some authors made to depend on the *day of the moon's age:* but as this is a very inaccurate mode of determining the moon's distance from the sun, the correct method is to make the correction depend on the difference of right ascension of the sun and moon; that is, on the *time of the moon's transit* expressed in *apparent* time.

4. It is found, however, that the correction does not depend on the difference of right ascensions of the sun and moon on the day of the observation, but at a certain antecedent time.   This time is antecedent by a different interval in different parts of the world, and we cannot as yet ascertain it for many places.   The rule given in the Annuaire du Bureau des Longitudes supposes the interval to be 36 hours, which is probably nearly true for the western coasts of Europe and the eastern coasts of America, but is not generally true.

5. The correction to be applied may be determined by means of the following table, which is calculated for the moon's mean parallax, and would require slight modifications for variations of parallax.

| Time of moon's transit at antecedent time | 0 | 1 | 2 | 3 | 4 | 5 | 6 | 7 | 8 | 9 | 10 | 11 | 12 hours. |
|---|---|---|---|---|---|---|---|---|---|---|---|---|---|
| Correction ......... | 0 | −16 | −31 | −41 | −44 | −31 | 0 | +31 | +44 | +41 | +31 | +16 | 0 minutes. |

The second line of numbers expresses the Semimenstrual Inequality of the interval of moon's transit and tide compared with the interval *corresponding* to the syzygy; and the correction of the observed interval is the excess of the semimenstrual inequality on the day of observation above the semimenstrual inequality on the day of syzygy.

Thus at London the tide corresponds to the distance of the sun and moon $2\frac{1}{2}$ days antecedent. Hence on the day of syzygy, the tide corresponds to a difference of right ascension of the sun and moon, amounting to 10 hours; for in $2\frac{1}{2}$ days the moon's motion in right ascension with regard to the sun is 2 hours. Therefore the tide-hour, on the day of new and full moon, is affected by the half-monthly inequality to the amount of 31 minutes additive. Suppose that when the moon's transit happens at $5^h$, it is observed that the tide is 44 minutes after the moon. The tide corresponds to a transit $2\frac{1}{2}$ days sooner, when the difference of right ascension of the sun and moon was $3^h$, and the half-monthly inequality for such a difference is 41 minutes subtractive. Therefore, in consequence of the half-monthly inequality, the tide interval on this day will be less than at the syzygy by $31 + 41$ minutes, and the tide interval on the day of syzygy would be $31 + 41 + 44$, or $1^h 56^m$, which is the vulgar establishment.

6. *The corrected establishment* is the establishment corrected for the half-monthly inequality; or the interval of the moon's transit and tide, not *on* the day of syzygy, but *corresponding to* the day of syzygy.

It may be determined by observing the intervals of the moon's transit and tide every day for a semilunation, and taking the mean of them. If several semilunations (a whole number, of course,) be observed, the result will be more accurate.

If we know by how much the transit of the moon to which the tide corresponds, is antecedent to the transit next preceding the tide, we may obtain the corrected establishment from an observation of any tide:—thus, in the above case, the tide being 44 minutes after the moon, and the half-monthly inequality 41 minutes, the corrected establishment is $1^h 25^m$.

It would simplify all our reasonings concerning tides, to employ in all cases the corrected instead of the vulgar establishment. The observation of the tides for a fortnight would give a first approximation to this; and observations continued for some months would give considerable accuracy.

### 3. *On the Effects of the Age of the Tide.*

The interval of time, by which the difference of right ascension of the sun and moon, to which the tide corresponds, is antecedent to the difference when the tide takes place, is, as has already been stated, what we have called the *age of the tide*, and has been named by Mr. Lubbock, following Laplace, the *retard*.

Its effect appears in two circumstances :—

First, In the difference of the corrected and vulgar establishment of any place. The interval of the moon's transit and tide at syzygy is greater than the mean interval for a fortnight; the excess is the increase of the half-monthly inequality during the age of the tide, reckoned from syzygy.

Second, The greatest and least tides do not happen on the days of new or full moon, but one, two, or three days afterwards.

As, however, we have attended principally to the times of high water, omitting for the present the discussion of the heights in detail, we shall consider only the effect of the age of the tide as it affects the times.

The exact determination of the age of the tide requires a considerable number of observations. When it is determined at one place for any derivative tide, we may expect that the same derivative tide at any other places will be deducible from the age so determined by considering the time which the tide-wave employs in passing over the interval. But when original tides may be supposed to come into play, it is not so obvious what will be the relation of the age of the tide at different places; for instance, at the Cape of Good Hope and at Van Diemen's Land. It would be very desirable to have continued observations at such places.

### 4. *On the Mode of reducing Tide-Observations.*

The best mode of obtaining from a considerable series of tide-observations, at the same place, the establishment and the age of the tide, (the elements which would enable us to construct tide tables for any place,) appears to be that employed by Mr. Lubbock and Mr. Dessiou, in their examination of the tides of the Port of London, and of other places. It is the following :—

The times of high water are arranged according to the half-hour of the moon's transit on the day of the tide:—thus, all those tides which took place when the moon passed the meridian between $0^h\ 0^m$ and $0^h\ 30^m$ (apparent time,) are put in one class; all those tides when the moon passed between $0^h\ 30^m$ and $1^h\ 0^m$ in another class; and so on.

The mean of all the times of transit in each class is taken, and the mean of the intervals of transit and high water. We have thus a series of times of transit, with the corresponding intervals of transit and high water.

The interpolation for other times is most easily performed by means of a curve, drawn on paper ruled into small squares. The times of transit being laid down as abscissas, the intervals of transit and high water are erected as ordinates, and a curve is drawn, approximating, as far as a regular form will allow, to the points thus found. This curve gives the intervals for *any* times of transit, and a table may be constructed by means of it.

The interval corresponding to the time of transit $0^h\ 0^m$ is the vulgar establishment   the mean interval is the corrected establishment.

The age of the moon when the interval is equal to the mean interval (the age being taken from the last syzygy,) is the age of the tide.

In this deduction, we neglect the inequalities produced by the variations of the declination and parallax of the moon and sun, which belong to a more advanced stage of the approximation, and have been treated of by other authors.

It may be observed that the difference between the greatest and least intervals of transit and tide is the same, for simple tides, at all places; it amounts, for the mean parallax and declination of the moon, to $1^h\ 28^m$.

We may observe also that the regularity of the tides is rather increased than diminished by their having to travel along a confined channel; the inequalities being extinguished by the narrowness of the passage, (as LAPLACE observes with respect to Brest,) like the oscillations of the mercury in the marine barometer. Nor does it appear that the irregularity of the channel much disturbs this result. The age of the tide might therefore be determined by means of observations made in an inlet or tide river. The establishment of the coast would not be thus obtained, but might be determined from comparatively few observations made at the mouth of the inlet or river.

5. *On tracing the Motion of the Tide-Wave by Comparative Observations.*

The places of the cotidal lines, and consequently the motion of the tide-wave, have been determined by means of such statements as we have been able to find, of the establishments of different places; and if the establishments were more correctly and extensively known, we might draw these lines with more accuracy and completeness.

But it is often possible to trace the motion of the tide-wave in a more compendious manner than by obtaining independently the exact establishment of many different but adjacent places. We may compare the times of high water at such places immediately with one another, and thus trace the motion of the tide-wave directly.

If, in travelling along any coast, we observe every day the time of high water, we can determine whether the establishment becomes later or earlier as we advance; for the effect of the half monthly establishment is known, at any rate approximately, if we know the age of the tide exactly; and by correcting for this we shall see how the establishments of different places are related to each other, and thus in what direction the tide-wave is moving.

This mode of observation would, however, be open to considerable inaccuracy, since accidental causes often accelerate or retard the time of high water by a quarter of an hour, or in some cases much more. This inaccuracy may in a great measure be remedied *by comparing each day's observation with those at a neighbouring place, where constant observations are made.* By this means the course of the tide-wave might be traced with considerable correctness by means of a single observation at each place.

This method of comparing the observed time of high water at any place with the time at some standard place in the neighbourhood, would be attended with great advantages. It resembles in principle the method adopted by astronomers of comparing the places of small stars and other celestial objects with that of some principal star in their neighbourhood.

By such observations as we have just spoken of, it would be easy to determine the points of divergence and the points of convergence of cotidal lines; that is, the places where the tide hour is earlier than it is at places adjacent on each side, and the places at which it is later. The determination of such points would be highly important for the purpose of enabling us to trace the course of the cotidal lines, and ought to be attended to in every maritime survey.

In a coast broken into bays, there would be many points of divergence and convergence on a small scale. In order to determine the general course of the tide-wave, it would be proper to make observations at points similarly situated; for instance, at all the principal promontories, or in the parts of the bays nearest the open sea.

It would also be proper to attend to the course of the flood and ebb streams, so as to determine from what quarter the tide comes according to the common notions on the subject. It is to be recollected, that this is not necessarily the quarter from which the tide-wave comes; but a knowledge of one fact may be of service in determining the other. In channels where the stream of flood and ebb run alternately in opposite directions, it is highly important that the time of *slack water*, as well as of high and low water, should be noted.

The publication of tide observations, as originally made, may be of the same use as the same proceeding in other departments of astronomy. It enables all, who are able and willing, to employ such observations in exemplifying, correcting, and extending the theory, and in verifying what is done by others in these ways. The publication of the Sheerness Tide Observations, recently directed by the Council of the Society, may thus be a useful step. It may be noticed that the Sheerness tides, like the rest of the Thames tides, are probably compound, being derived partly from the north and partly from the south branch of the British tide. On this account, some of the circumstances may be disguised in those tides and in the London ones, which would appear in simple tides; for instance, the difference of the two semidiurnal tides. The tides of Plymouth, Falmouth, Penzance, or some point in the Scilly Isles, would in this respect be more instructive, and would offer an interesting comparison with those of Brest.

### *Conclusion.*

I cannot conclude this memoir without again expressing my entire conviction of its very imperfect character.—I should regret its publication if I supposed it likely that any intelligent person could consider it otherwise than as an attempt to combine such information as we have, and to point out the want and the use of more.   I shall be neither surprised nor mortified if the lines which I have drawn shall turn out to be in many instances widely erroneous : I offer them only as the simplest mode which I can now discover of grouping the facts which we possess.   The lines which occupy the Atlantic, and those which are near the coasts of Europe, appear to have the greatest degree of probability.   The tides on the coasts of New Zealand and New Holland have also a consistency which makes them very probable.   The Indian Ocean is less certain ; though it is not easy to see how the course of the lines can be very widely different from that which we have taken.   The course of the lines in the Pacific appears to be altogether problematical ; and though those which are drawn in the neighbourhood of the west coast of America connect most of the best observations, they can hardly be considered as more than conjecture :—in the middle of the Pacific I have not even ventured to conjecture.

It only remains to add, that I shall be most glad to profit by every opportunity of improving this Map, and will endeavour to employ for this purpose any information with which I may be supplied.

Good tide-observations, at almost any place, will be valuable for this purpose, if the local circumstances be known.   The following may be mentioned as instances in which such observations would be of more peculiar value.

1. *Good observations* at the Cape of Good Hope, at Van Diemen's Land, at Swan River, and at some place on the east coast of New Zealand : also at any place on the south or east coast of Ceylon ; at the Mauritius, and generally at any of the islands in the Indian Ocean.   Such observations would serve to decide the general position of the cotidal lines.

We want also observations on the west coast of America and at some of the islands of the Pacific, where the tides are sufficiently large to be clearly distinguishable from a rise or fall of the surface of the sea due to other causes : but such observations must be made at several places before they can be connected in any very probable manner.

2. *Good and long-continued observations* at any of the above places, or at other places, whether in an open part of the ocean, or not : for instance, those made within the inbend of deep sounds and bays, or on the banks of rivers, would be no less valuable than those on an open coast, or detached island, for certain purposes.   These observations would enable us to determine the

age of the tide, the difference of the two semidiurnal tides, the effect of the parallax and declination of the moon, and many other circumstances of great interest.

3. *Comparative simultaneous tide-observations* at different places on the same line of coast.  These would enable us better than any other observations to determine the motion of the tide-wave along the coast.

Probably tide-observations at all the points of the coast of England where the officers of the Preventive Service are stationed, carefully made and continued for a fortnight, would give us a clearer view of the progress of the tide along our coasts than we can obtain by any means at present extant.

In surveying any coast, one part of the proceeding ought to be to station an observer at one point to observe the tides constantly, and to employ others to move from place to place, noting the tides at each, so that they may afterwards be compared with those at the tide-station.

---

*₊* This Paper is accompanied by two Charts;—namely, A general one, representing the greater part of the world, and A Chart of the British Isles; on which THE COTIDAL LINES are drawn, according to the observations adduced and discussed in the preceding Memoir.

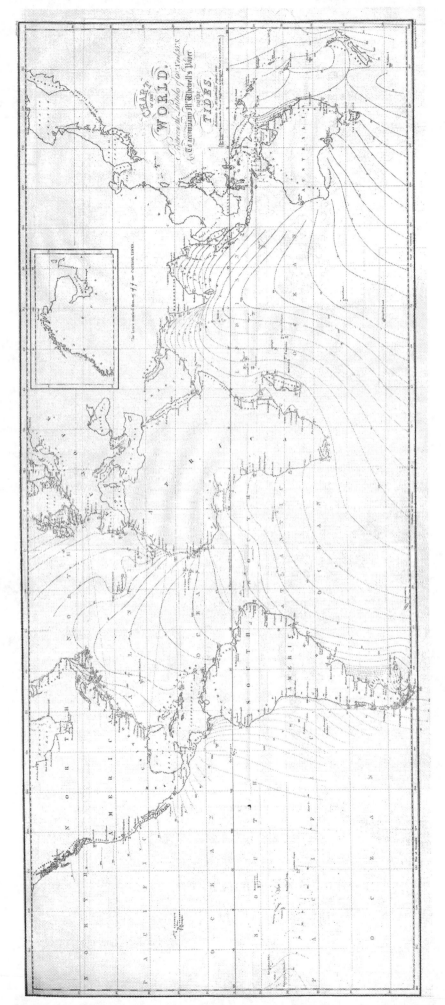

The material originally positioned here is too large for reproduction in this
reissue. A PDF can be downloaded from the web address given on page iv
of this book, by clicking on 'Resources Available'.

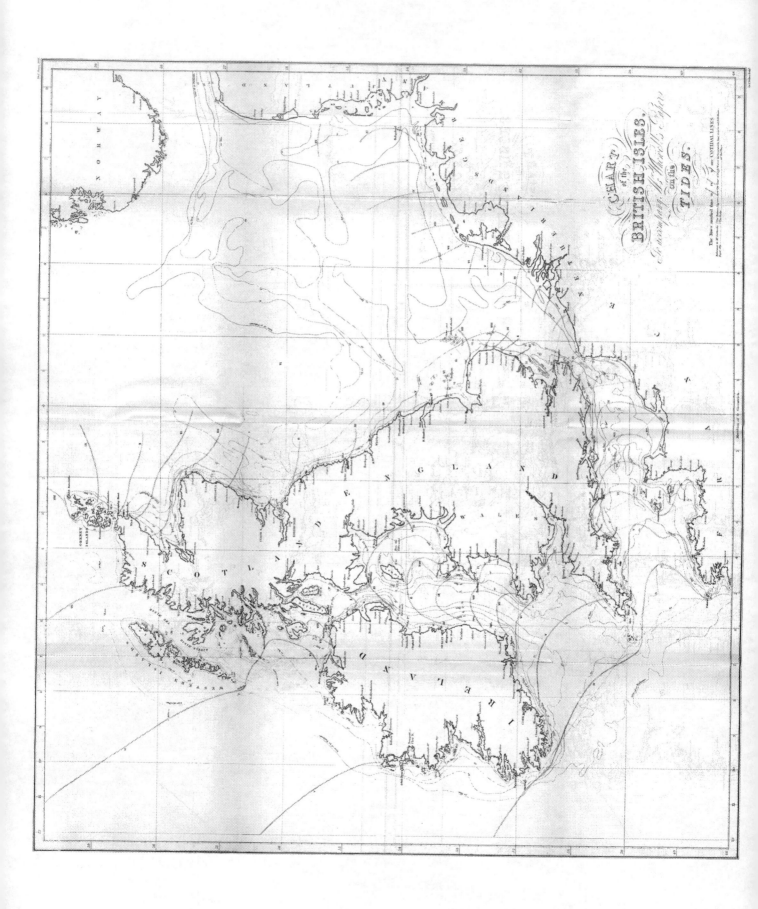

Printed in the United States
By Bookmasters